Many Worlds to Conquer

TJ Halbertsma

authorHOUSE

AuthorHouse™ UK
1663 Liberty Drive
Bloomington, IN 47403 USA
www.authorhouse.co.uk
Phone: 0800 047 8203 (Domestic TFN)
         +44 1908 723714 (International)

© 2019 TJ Halbertsma. All rights reserved.

No part of this book may be reproduced, stored in a retrieval system, or transmitted by any means without the written permission of the author.

Published by AuthorHouse 12/26/2019

ISBN: 978-1-7283-8593-8 (sc)
ISBN: 978-1-7283-8594-5 (hc)
ISBN: 978-1-7283-8592-1 (e)

Print information available on the last page.

Any people depicted in stock imagery provided by Getty Images are models, and such images are being used for illustrative purposes only.
Certain stock imagery © Getty Images.

This book is printed on acid-free paper.

Because of the dynamic nature of the Internet, any web addresses or links contained in this book may have changed since publication and may no longer be valid. The views expressed in this work are solely those of the author and do not necessarily reflect the views of the publisher, and the publisher hereby disclaims any responsibility for them.

Dare to dream and make it happen!

For my nephew Felix

And in memory of Marc Cornelissen
and Rémy Lécluse

Foreword

I first met TJ in a corporate setting, and then, to my surprise, I next saw him in the Antarctic, where I learned of his various adventures. Having seen him in action in these different environments, I know that TJ is a guy who makes things happen. He shows us all that if you get out of your comfort zone, you will go far. He also cares for our world, which is why I salute TJ not only as a fellow adventurer but also as my friend!

—Robert Swan OBE, the first person to walk to the North and South Poles

Contents

Introduction xiii
Chapter 1   Ain't No Stoppin' Us Now 1
Chapter 2   Falling from the Sky.................................... 18
Chapter 3   Still Aiming High 31
Chapter 4   Peak Performance....................................... 47
Chapter 5   Completing the Trilogy............................... 65
Chapter 6   The High Life Continues............................. 79
Chapter 7   On Top of the World.................................. 100
Chapter 8   This One's for Marc................................... 130
Chapter 9   Escape from Alcatraz................................. 155
Chapter 10 Crossing the English Channel 171
Ain't No Mountain High Enough 182

Introduction

"Mum, would it be OK if I cycle to school this afternoon?"

"Definitely not. You're five years old."

"But Mum, my training wheels have just come off. I know how to ride a bike."

"Don't be ridiculous. Just finish your lunch, after which I'll drive you back to school. And please hurry up, as you don't want to be late!" My mother left the kitchen to get her car keys, and when she returned only moments later, I was no longer there. She asked my twin brother where I had gone, and he shrugged his shoulders. My mother searched the house but couldn't find me anywhere. Then she realized: Elvis had left the building …

She jumped into the car with my brother and told him to look out for me on my bike. I remember very vividly that I was enjoying the freedom of the roads, but also that those passing cars on the Prinsenweg in Wassenaar are a lot scarier when you're on your little bike than when you're in the back of your mother's car. She couldn't find me on the road, as I had made a stop to see Annemiek, the prettiest girl in class who lived on the way to school. Her father was surprised to see me when I rang the doorbell. "TJ, what are you doing here?"

"I'm here to see Annemiek."

"Well, she's back at school already. Shouldn't you be there too?"

"Yes, sir," I replied. "I'll see her there then!" He was baffled and tried to stop me from continuing, but I rushed off on my bike to finish my adventurous journey to school.

I was hoping for a hero's welcome when I made it to the schoolyard, as I hadn't fallen once during those 4 kilometres! But a hero's welcome was not what I got. My mother and my teacher were waiting for me at school, absolutely furious. I had to promise I would never do such a stupid thing again. I pretty much knew it had been foolish indeed as I had had a couple of near misses with those cars rushing by. Even though I was only 5 years old, I realized that the dangers had been far greater than the reward of making it to school on a bicycle without training wheels. I had been lucky. It was an early lesson that one has to carefully weigh risk and reward. I guess you could also say that little adventure was the start of my exploring career.

That question of risk and reward always comes up when I talk about my mountaineering adventures. Many people ask me, "Why would anyone risk their life climbing a mountain?" Sure, it makes you feel alive and gives you a great sense of achievement to stand on a summit. But you put yourself in danger, which is hard to justify to anyone. And you don't do it to explore uncharted territory, as every mountain has been climbed already. But people still do it, including me. And I absolutely love it. Perhaps the best answer came from the legendary George Mallory when he was asked in 1924 by a *New York Times* reporter why he was planning to climb Mount Everest. He famously replied, "Because it's there." Try to argue with that.

When I was in Turkey in 2014, I got a different perspective on this subject. I was there to swim the Hellespont, which is the great waterway that connects Europe to Asia. This swim can only be done once a year, when the Turkish authorities close the strait to all boat traffic. That normally happens on 30 August, which is Turkish Victory Day. But the event was not held on 30 August that particular year because the waves were too high and the currents too strong for any swimmer to make it safely to the other side. The event was postponed for two days, and all I could do was hope for better conditions. It also gave me the opportunity to meet some other swimmers who were going to attempt this legendary swim from Europe to Asia.

In the hotel in Çanakkale I had the pleasure of meeting a gentleman called Andrew Rice and his wife Hazel, who was also planning to swim the Hellespont. They were both retired. Andrew told me he wasn't a swimmer but said he was there to support his wife in fulfilling her lifelong dream of swimming the Hellespont. They were working on their bucket list.

It didn't take long before Andrew and I started swapping war stories of our adventures. He told me about his successful climb of Mount Vinson on Antarctica, after which I told him about my Last Degree expedition to the North Pole. He said, "Wow, TJ, you've done things that people would love to hear about. Would you be interested in speaking at the Allenburys Adventure Series?" I had never heard of the Allenburys Adventure Series, but he told me it is a monthly event he organizes at the GSK Sports and Social Club, just outside of London. It would be easy for me to get to, he said, and he added that many prominent adventurers had spoken there before. I simply had to laugh and politely declined.

I told him, "Andrew, I'm not a real adventurer. I have a demanding job in London, and I go to an office every day in a suit!" I added that I only did a bit of adventuring on the weekends and during my holidays.

He shook his head and said, "TJ, that is the whole point. People would love to hear your stories because you're a regular guy. They can relate to you! It's inspiring!"

Although I was still not convinced that people would be keen to hear about my adventures, I accepted his invitation to present on my North Pole expedition and was due to speak on 9 December that year. When I read the flyer, I saw that people would be required to pay to attend the evening. It was only £2 and aimed to cover the room rent, not to pay me, but I was excited. People would pay to hear my stories. Wow! But would anyone really show up?

Andrew had been right. Almost a hundred people showed up, and they absolutely loved it. The crowd wanted to know everything about my expedition: where I had gotten the idea, how I prepared, whether or not I'd seen a polar bear, and interestingly, if they could do this too.

Only then did I realize that my adventures could have a purpose after all. If I can inspire others with my stories to get them out of their comfort zone, to get them to do things they didn't think they could, then that really makes every bit of suffering on my adventures more than worth it.

That evening also encouraged me to write down ten of my adventures that I have collected here. I hope you will enjoy the stories and that they will inspire you. And remember: I'm a regular guy who goes to the office every day, so if I can do this, you can do this too. In fact, you can achieve anything if you just go for it!

Chapter 1

Ain't No Stoppin' Us Now

I grew up in the Netherlands, in a suburb of The Hague. My early childhood memories are all about fun: birthday parties in the garden, jumping in the swimming pool, and playing hide-and-seek in the big house. That unencumbered life changed rapidly when my parents divorced. My mother, twin brother and I moved into a little flat in The Hague. My brother and I were six years old, my mother didn't have a job, and the contrast between our old life and our new life couldn't have been greater. No more space at home, definitely no more garden or swimming pool, unfriendly neighbours, and a very long drive to school—definitely too long to ever attempt on a bike!

To leave my father without any financial security and with two young boys to take care of had been a big decision for my mother. The day we moved into our new place, my mother wanted to unpack the boxes. She sat me and my brother down at the one table we had. She wanted to keep us busy, so she gave us some paper and felt-tip pens. She said to go and draw, and then she started opening the boxes. After half an hour, she checked on us and said, "What are you drawing?"

"Take a look, Mum." I proudly showed her the clown I had drawn.

"But he looks so happy," she said with a trembling voice, "with flowers in his hat and a big smile on his face!"

"Of course, he's a happy clown," I replied, not realizing that my mother had been very apprehensive of how we would cope living in this little flat without my father.

Years later, my mother told me it was there and then that she realized she had made the right decision. I greatly admire my mother for having had the guts to make that decision. Despite the fact that my mother had to get a job and we had to live on a small income during those years, my brother and I had a very happy childhood.

The good thing my parents' divorce taught me is that money and success can be taken away from you in an instant, so you should never take anything for granted. That gave me the drive, at a very early age, to work hard and push myself.

As a young kid I found sports a good way to express myself and show who I was. There was a football club next to our block of flats. I went there on a Saturday morning to check it out. That was not a great success. Three kids immediately came up to me and asked me what I was doing there. "Well," I said, "I live in that building and just wanted to check out this football club."

"We don't want you here. Get out," I was told. Clearly the kids in my new neighbourhood were not as friendly as I had hoped for. Fortunately my mother realized she had to find a fun sport for my brother and me to blow off some steam, and she encouraged us to play field hockey, which is one of the most popular sports in the Netherlands. She enrolled us in a club, and hockey quickly became my passion. I loved the pace of the game and being part of a team. And what I lacked in technique, I made up for in effort and speed. I was proud to be selected for the first teams of my age group at my club, the Kieviten.

My big break came years later when I was 15 years old. We had moved to a bigger place closer to the hockey club, which had given me the opportunity to spend a lot of time on the field to train. My coach had said that he would propose me for the regional team, which was a big surprise for me as not many kids from our club were ever put forward. The Kieviten was more about having fun on the field and less about top-league hockey. And the only kids who were ever put forward were those whose fathers did a lot for the club. My dad was certainly not in that category. He was a lot older than the other fathers and wasn't very sporty. I remember the one time he did come to the club to watch me play, but he arrived late and was chatting to all the other parents rather than watching me play. Afterwards he didn't even know which team had won!

The fact that I was still proposed for the regional team meant that it was based purely on merit, which boosted my confidence. There would be two selection rounds. The first one would be at our club, where I obviously had a home advantage. I was a defender, which makes it difficult to get noticed as you don't score goals; you just have to prevent the opposition from doing so.

Just before the game started, I talked to a friend from Premier League club HDM who would also be playing in this selection round, and he gave me very sound advice: "TJ, you're a defender. You play for a minor club. No one knows who you are, so make sure you're as conspicuous as possible. And definitely volunteer to take a penalty stroke when your team is given one."

The first half was pretty uneventful, and I didn't get many opportunities to be in the picture. But I couldn't believe my luck in the second half. After a deliberate infringement from the other team in their penalty circle, our team was given a penalty stroke, but no one wanted to take it. This was my chance!

I ran across the field and shouted, "I'll be happy to take it!" The others were pleased to pass the hot potato to me, as a miss would immediately end a player's hopes of being selected. For me this was a much-needed opportunity to get myself noticed. No one objected when I volunteered. The next thing I knew, I was walking to the penalty spot with everyone watching. The pressure was on, but I knew I had to take this chance. Defenders are not natural goal scorers, but I always felt confident to flick the ball hard, bottom left, and that's exactly what I was going to do now. I didn't want to let the keeper distract me. I looked down and focused on the ball, just waiting for the umpire to blow his whistle. When he did, I flicked the ball hard, bottom left, as planned, but the goalkeeper dove in the right direction. Fortunately he was too late. The sound of the ball hitting the wooden bottom board was music to my ears. *Goal!*

I was incredibly relieved and found it very hard to contain my excitement, but I also knew it would be better to keep my cool when I walked back to our half of the field. *Make it look like business as usual,* I thought. After the game, Marc Delissen came up to me. I couldn't believe it! Marc Delissen not only was heading the selection committee but also was the best player in the Dutch national team, a total legend. I was a bit intimidated, but he was very friendly and asked my name. I told him who I was and said that I was very grateful for the opportunity to play in this selection round. He said, "Well, TJ, I look forward to seeing you again in two weeks' time, as you're through to the second round." Not only had I just talked to the best player in the country, but also I was one step closer to making it to the regional team!

The second round took place at Hockey Club Forescate in Voorschoten, where I had also played before. I made sure not to do anything stupid during this round. I played well but hadn't had a similar opportunity as in the first round to shine, so when the game was over I was pretty anxious to discover whether I actually had made the cut. Half an hour later the selection would be announced, and there were still four players for each available position. When the team was finally announced, I had almost given up hope when all the names of the usual suspects from the top clubs were mentioned, but then my name was called as one of the last ones. I had made it to the regional team after all!

During that year in the regional selection I learned a lot and was very honoured to be part of that team of outstanding players, which lifted my game and also gave me the confidence to move to a better club. I had thoroughly enjoyed my time at the Kieviten and had made lots of friends there, but if I wanted to progress in my hockey career, I needed to move on.

The following year I moved to HGC, the Premier League club that Marc Delissen was also playing for. I had two great years there and played hockey at the highest level in the country. But when I went to university, I had to give up my hockey career. Those experiences on the field were very formative as I had learned that with hard work and dedication, one can go a long way.

After university, I took time out to study French at the Sorbonne in Paris. I had a great time, met many interesting people, and was introduced to Marco by a mutual friend. He was a good-looking guy who worked as a music executive for Arcade Records. He was a couple of years older than I was, had beautiful girlfriends, was hanging out with pop stars, and was clearly living the dream. But what impressed me most was a picture in his apartment of him finishing the New York City Marathon. Mind you, this was 1994, and running a marathon was still seen as a real test of human endurance, the ultimate challenge.

Legend has it that the first marathon was run in 490 BC by a Greek man called Pheidippides. He is said to have run 42,195 metres from the battlefield of Marathon to Athens to announce victory over the Persians. Upon delivering the good news, he collapsed and died of exhaustion. So, to participate in a marathon was something I very much assumed to be for professional athletes only and maybe some hard-core amateurs.

Nowadays, marathons attract many different types of participants, including many recreational runners. But in the early days, it was a different story. In 1976, the first year the New York City Marathon was held, about 20 per cent of runners broke the three-hour level. Today, that number is only 2.5 per cent, which shows that a marathon has become a playing field for anyone looking for a challenge.

Marco was the first person I'd ever met who had run the New York City Marathon. When I asked him about it, he said, "Anyone can do it as long as you're willing to train hard and endure pain during those gruelling forty-two kilometres." I asked him whether he thought I would be able to do it too. He said, "You've been fit all your life, playing hockey and then rugby at university. You would have to train for it, but you're definitely able to do it." That sounded very encouraging and also exciting. And I had never even been to New York.

After Paris, I had to look for a job. When I was still at university, I had been lucky enough to secure an internship in London in 1993, where I had immediately felt at home. The dreams, the drive, the ambition of the people I met there—I just loved it all. I knew I wanted to go back there to start my career, so after Paris I applied for a job in London and managed to find a good position at a big financial services firm in 1995.

It was a year later, in the autumn of 1996, when I made my first trip to New York. I had saved a bit of money and went for a long weekend. I had never been to the United States, and my visit to New York was truly overwhelming. The can-do attitude of the people and the idea that all your dreams can come true really appealed to me. Anyone who's been to New York knows it is a city with so much energy, so much positive attitude, so much draw; it was an amazing experience in this city that never sleeps. The roasted chestnuts that vendors were selling on every street corner smelled delicious, and the skyscrapers were bigger than I had ever imagined. The Brooklyn and Harlem signs made me proud as those names are derived from Breukelen and Haarlem, two towns in the Netherlands, and stem from the days when New York was still a Dutch settlement called New Amsterdam. It was bitterly cold, and hot steam was blowing from every manhole cover on the street, with the iconic yellow taxis whizzing up and down the avenues. It also happened to be the weekend of Halloween, and everyone I saw in the street was dressed up in spooky costumes going to fancy parties. In short, New York looked like a movie set to me—unbelievable. My introduction to New York was similar to when I visited London for the first time in that it opened a new world to me and I wanted to have a piece of it. All of a sudden, that picture of Marco running the New York City Marathon came to mind. This was it; I had to do that too. It would be my way of taking New York!

Back in London, I discussed running the New York City Marathon with Adrian, a delightful guy who is always up for a challenge. He was introduced to me by a mutual friend he was dating, and she had told me he was a keen runner but that he wasn't always as responsible as he should be. It would soon became clear to me why she'd said this.

Adrian didn't need much convincing when I suggested the New York City Marathon to him. "Yes, that sounds amazing. I'm definitely in. We should immediately start planning. There's one thing I have to tell you: I have type 1 diabetes."

"OK, what does that mean?" I asked him.

"It's an autoimmune disease that prevents the pancreas from producing insulin. So I need to inject myself with insulin multiple times a day." I told him I had heard of diabetes but thought it was a condition mainly affecting older obese people, a category which Adrian certainly didn't fall into. He explained that I was referring to type 2 diabetes, which is very different from type 1. "Type 2 can often be treated with more regular exercise and a better diet. Unfortunately, type 1 cannot be treated in a similar way." He continued, "And indeed, type 2 is generally diagnosed in adults, whereas type 1 generally affects adolescents." He finished with something I didn't like: "Most importantly, type 1 is much more dangerous."

This got me slightly worried, and I asked him whether he should be running a marathon then. "Not a problem. I know what I'm doing, so you don't need to worry about anything." That was good enough for me. After all, he was a sensible person, Cambridge educated with a serious career and fit as a fiddle.

Adrian and I both had busy jobs, and we didn't get the opportunity to train together. A couple of weeks after we both had started to train—it was the spring of 1997—I got a call from his girlfriend. She had found Adrian unconscious in his front garden after a running session. She told me, "I saved his life by getting him instant emergency care at the hospital. But if I hadn't found him, he would be dead. And this is not the first time this has happened." Clearly Adrian was not taking his condition seriously enough. She was angry and upset. "Could you please promise me to take good care of him during the marathon? I simply can't rely on him to be responsible and take appropriate care of himself." I promised her I would, but this certainly added an extra dimension to the challenge of running the New York City Marathon.

My training didn't have a smooth start either. I was given a book that was written by the running coach Wim Verhoorn which gave me some basic training tips. In short, you have to gradually build up your running and make sure not to overdo it so you avoid injuries. At the end of your training regime, you finish with half a marathon. Then you start to taper a couple of weeks before the event to save your energy.

When we received confirmation in early spring that we had been allocated a starting place for the 1997 New York City Marathon, I decided to up my training regime by running farther distances. I was so excited that I started to train hard and would go well beyond the 10 kilometres I would normally run. I should have known better; I ended up with an inflamed right knee. I had never had this before, and it was a clear sign I should have increased my training more gradually. The physiotherapist who treated my knee said I had to take a couple weeks of rest—no running—and that I should forget about running a marathon. *Really? We'll see about that!* After my three weeks of rest, I slowly started my training regime again. I was fortunate to still have a couple of months before the actual marathon, which would take place in November. I managed to gradually restart my training, mostly in Hyde Park, and every time I was finished, I felt amazing. It was not only the beautiful green space I would enjoy in the park but also the natural reward that running gives you. The physical exercise makes your body release endorphins, which induce a state of euphoria often described as a "runner's high".

A month before we were to leave for New York, Adrian and I came together to discuss the last details. An important feature, we had been told, was a shirt with your name on it. New Yorkers will shout your name if you need the encouragement—but probably also if you don't—and it really gets you going. Adrian didn't think it would be necessary, but I thought, *I'm going to use all the help I can get.* I printed my name on the back of my shirt as I knew I would have to wear my number on the front. He and I also agreed to run together, which would allow me to keep an eye on him. But we also knew it would be very likely we would lose each other among thirty thousand other runners.

In the two weeks before we left, I collected donations because I wanted to run the marathon for a charity. There really wasn't a particular charity that appealed to me, but I had heard of one that provides a "home away from home" for parents whose children are being treated in hospital. The children angle resonated with me, and I managed to raise some funds for the Ronald McDonald House charity. This also increased the pressure on me because I now had to make sure I completed the marathon. I actually wasn't too worried about not finishing the race, but everyone would now be asking me how I had done. So I thought I'd better perform well. I set myself a sub-four-hour goal. I knew it was very doable, but it still provided a challenge.

We left a couple of days before the event so we'd have plenty of time to get over our jet lag and get ready for the race. Adrian is a couple of years older than me and knew New York well. He suggested we go for dinner in an area in Lower Manhattan called Little Italy and that we do the same every night. "Well, Adrian, I also love Italian food, but do we need to do the same thing every night?"

He then said, "We need to have pasta every night. It's time to start loading up on carbohydrates because they give you energy. The more you have in your body, the better." I had heard that you have to be careful with carbohydrates when you have diabetes, as your body doesn't produce the insulin to break the carbs down. But we also had a marathon to run, and Adrian seemed to be coping well, even after three big pasta nights in Little Italy. I also felt good, as I had quit smoking for six months already, had trained hard, and had not had alcohol for a couple of weeks. I now wanted to get the marathon over and done with. I was ready!

The night before the marathon, Adrian managed to get us invited to a Halloween party. We thought it would be a good distraction, so we decided to go. We didn't make it late, we didn't drink any alcohol, and we weren't dressed for the theme, although I knew from my first visit to New York City that everyone dresses up for Halloween. Because of the last-minute nature of the invitation, we hadn't brought a costume, and as a result we stuck out like sore thumbs among this glitzy New York crowd. But the party was a good way for us to relax, and we were fortunate to meet Paul McKenna there, the positivity guru who teaches people that anything is possible as long as you believe in it. We couldn't have thought of a better person to put us in the right frame of mind the night before our big event!

Adrian and I woke up early that Sunday morning, 2 November, as we had to start with a bus ride from Manhattan to Staten Island, where we had to wait for a couple of hours before all thirty thousand runners had gathered at the start. It was a very long wait, but after three hours it was really going to happen. My starting number was 10864, so when I finally got to the starting line, more than ten thousand people were already in front of me and about ten minutes had already elapsed. This delay would not be accounted for at the finish, which I had not realized before the race. We were already ten minutes down without having run a metre! Nowadays runners have a chip so their time only starts the moment they pass the starting line, but such was not the case in 1997. Too bad. But we had a marathon to run, so we needed to get going!

You start the marathon by crossing the Verrazano-Narrows Bridge. I was glad I was finally moving as it was a good way to get rid of the nerves. I wouldn't say we were running; it was more like a slow jog as there were just too many people trying to cross this bridge at the same time. But after a couple of minutes we were descending the bridge. It opened up a bit, and we started to get into our rhythm. This was it; we were now actually running the New York City Marathon!

After the bridge, the course takes you through Brooklyn, firstly along Fourth Avenue. You get a good feel here of what makes the New York City Marathon so special as the crowd is already cheering you on. Many spectators were also playing loud music from their windows, and hearing "Ain't No Stopping Us Now" from McFadden & Whitehead made me feel truly invincible. It was just what I needed as the weather was turning bad: humid and windy, and it had just started drizzling.

Adrian and I were running together, and we both felt strong for the first hour. We were making good progress but also realized that we shouldn't make the classic mistake of giving it too much in the beginning only to "hit the wall" later on, which usually occurs around the 30 kilometres mark. Your body is tired and running low on glycogen, which is a carbohydrate that gives you the energy to keep running. Your body tells you to stop, but you're not there yet. We just had to remember that this was a marathon, not a sprint, and we were hoping we would avoid hitting the wall. But we were doing well so far, pacing ourselves. When we got to the Pulaski Bridge after 20 kilometres, which is the halfway point of the race, we realized we were on track for a three-and-a-half-hour finish. I didn't feel out of breath, I didn't feel any lactic acid build-up in my legs, and I'd been able to quench my thirst at every drinking station. So far, so good. *But we shouldn't get too excited*, I told myself. *Just stay focused and keep the pace.*

We ran through Queens, where we didn't see many spectators, but when we were running across Queensboro Bridge, we could already hear the crowds in Manhattan, which was very uplifting. Despite the roars, this is a point where many runners are beginning to feel it as they run up that bridge. Up until now Adrian had been absolutely fine, and I hadn't even thought about his diabetes. Running up the bridge, he said he was getting a bit weak and told me to go on if I wanted to. He had slowed down, was avoiding eye contact with me, and started to look pale, all of which I didn't take as positive signs.

I knew the next drinking station was not far away. I told him, "I can see we're close to the next drinking stop. You'll feel much better after a Gatorade energy drink." He didn't respond to that, but I assumed he'd be fine again soon and we would be able to continue running together at a normal pace.

When we got there he said, "I'm OK again. I don't need to refuel, but you go ahead. I'll catch up with you later."

I wasn't going to agree to that. I said, "Adrian, you will not be fine unless you drink. You need sugar to keep up your blood glucose levels." He had told me before the race that I had to make him drink something sugary if he got weak, so I was very surprised that he didn't want to drink to fuel up.

After I had finished my Gatorade, he started running again, so I assumed he was all fine again, despite not having had anything to drink. We were now at 26 kilometres and had made good progress, but we still had another 16 kilometres to go. And at this point it started to rain heavily.

We proceeded on First Avenue before we got to the Bronx, and Adrian started struggling again. Something wasn't right. He wasn't responding to me anymore. He had stopped running altogether and could barely walk. He kept on insisting that I carry on, but by this point I was getting worried about him. Because Adrian had told me before the race to make him drink if he got really weak, I decided to force him to drink the Gatorade at the next drinking station. This was not going to be easy, as he had refused to drink anything at the last station. When your blood glucose gets to dangerously low levels, your brain stops functioning properly before you pass out. That is called a hypo and is potentially fatal. Obviously I didn't know the level of his blood sugar, nor that he had injected insulin before the race, which had already lowered his blood sugar just like running the last 28 kilometres had. He seemed very close to passing out as he started to wobble and was no longer communicating with me.

When we got to the next drinking station, I was very clear: "You're not taking one more step unless you drink a full cup of Gatorade." I shoved one in his hand, and he was just standing there, doing nothing. "Unless you empty that cup, we're staying right here!" That wasn't a compelling thought as the rain was pouring down, we were both cold, and the clock was ticking. But that was the least of my worries at this stage. Adrian realized I was serious—we wouldn't be moving—and was too weak to resist again, so he ultimately drank his Gatorade. I was surprised how quickly he became himself again after he had finished his cup. Obviously it had worked! He was still slow, but at least he could jog again. And he started to speak again. I made sure we were going to stop at every drinking station we encountered until the finish, whether he liked it or not!

By now I was starting to feel it too. My legs just didn't want to go on anymore. It was clear this was the wall that everyone had talked about. You still have about 10 kilometres to go, and your mind tells your body to stop any form of activity so as to preserve your energy. That combination of your body and your mind telling you to stop makes it doubly challenging to keep going. But you have to push yourself through this. The cheering crowds definitely helped. I could only imagine how much Adrian was suffering, but he kept going, albeit slowly. After we proceeded through Harlem, we ran down Fifth Avenue into Central Park, where people were shouting out my name, which they saw on my shirt. I knew we were getting close, and I hardly noticed the puddles of water we were running through.

Coming into Central Park was psychologically important as Adrian and I both knew that the finish was not too far now, only 4 more kilometres to go. But our optimism was short-lived as we soon encountered the hills which made our legs hurt even more. I drew a lot of strength from the crowds who were still chanting my name and kept shouting, "Doing great." Obviously I wasn't, but it certainly helped me a great deal! It also forced me to look up to thank those who were calling out my name. I drew a lot of energy from their encouragements, but Adrian didn't have his name on his shirt, which proved to be a disadvantage as no one was calling his name.

I shouted at the crowds and asked them, "How's Adrian doing?"

The immediate response was "Doing great, Adrian," which put a big smile on his face. Knowing his agony was almost over, he had to draw on every reserve he had left in his body to make it to the finish line.

I had kept track of the time and realized that we had slowed down significantly since the halfway point. I was hoping we would still finish within four hours, but when we got to Columbus Circle, it had become clear that that would be very difficult.

The last kilometre was very tough, but it was exciting at the same time as I started to take it all in. We were about to take New York, and nothing would be stopping us now, whether it was the rain, our exhaustion, or diabetes. I also thought of the thousands of excited spectators who must have been very cold, standing in the pouring rain just to cheer us on. I can't tell you how much I appreciated it as it really lifted our spirits. We slowed down when we were 30 metres away from the finish as a large group was in front of us. We knew a photo would be taken of every runner going through the finish, and we wanted the photo to clearly show it was us.

It was a great moment for both of us when we passed the finishing line at 4.00.53. I gave Adrian a big hug. We had lost ten minutes at the start, so one could argue that we had run the marathon in less than four hours, but chip timing simply didn't exist in 1997, so we just had to accept that we hadn't broken the four-hour mark.

Supporting Adrian across the finishing line. We did it!

But that was a minor issue. What was important for me was that I now knew I could do things that I had not thought were within my reach. For Adrian, this had also been a life-changing event. Not knowing what he was physically capable of as a diabetic was a question he now had an answer to. He didn't have to be held back by diabetes; he could do what others were able to do, as long as he managed his condition. Without knowing it at the time, this was a valuable lesson for me too, as diabetes would become an important part of my life later on.

Chapter 2

FALLING FROM THE SKY

After our return from the New York City Marathon, Adrian and I immediately started contemplating what our next challenge would be. An obvious one would be the London Marathon the following year, but to be fair, I was looking for something else. I had now proven to myself I could run a marathon, so I didn't see the point of doing another one. Then Adrian came up with an idea that would be very different indeed and would certainly get me out of my comfort zone. "OK, TJ, if you're looking for something thrilling, something that gives you the ultimate rush, why don't we go jumping out of planes, also known as skydiving?" Adrian had done a few jumps already, so he knew what he was talking about.

I was a bit hesitant. "Adrian, I've never even considered bungee jumping, and now you're suggesting we go skydiving? I like to push myself, but this sounds like total madness!"

He then went on to explain what skydiving is about. "It is not the same as parachute jumping, where you jump out of an aircraft at fifteen hundred metres and the static line parachute will open automatically as you leave the plane." This is actually what I had been hoping for, but it was about to get even scarier. "Skydiving really is the next level and the purest way to enjoy the freedom of the skies. You will jump out of an airplane at four thousand metres in free fall until you reach a speed of one hundred eighty kilometres, after which you will have to open your parachute yourself and find a safe spot to land." This didn't sound appealing to me at all.

Unlike Adrian, I hadn't done any static line parachuting before, so I wasn't sure I was ready for skydiving, but then I was assured by Adrian that an advanced free-fall (AFF) course would make me a skydiver in no time. An AFF course is an intensive course aimed at people who have never jumped out of a plane before, and it will progress you four times faster than the conventional static line system where you're tied to an instructor and the parachute opens the moment you jump out of the plane.

I have to admit that the idea of falling from the sky did appeal to me, but at the same time it sounded very frightening. I mean, if something goes wrong with your parachute, you're very likely to fall to your death. And if all goes well, then you've had a good adrenaline rush. I wasn't convinced that this was something I had to do. What I did like about this potential adventure is that it did not involve endurance, and that meant that Adrian's diabetes would not be an issue.

I told Adrian I would think about it, and he felt that I wasn't convinced. He then invited me to come along to Netheravon, which is an army airfield and the number one spot for skydiving in the United Kingdom, operated by the Army Parachute Association. Adrian introduced me to Andy Parkin, who had become a full-time skydiving instructor after his army career. Andy explained to me what the AFF course would be all about. "It consists of eight levels or jumps, and for the first three jumps you will be accompanied by two instructors. For the next four jumps you will only have one instructor, and on the last jump you're on your own. It's great fun. You'll love it, TJ!" Andy was a really nice guy. He had great energy and made it all sound like fun. He had a very good sense of humour, and we laughed a lot, which very much helped to settle my nerves. He saw I was warming up to the idea. "Would you like to try a jump now?" Andy asked. I wasn't ready mentally and declined, but I had indeed started to seriously consider skydiving. He subsequently told me it would be best to follow an AFF course anyway in a country where the weather is more predictable than that in the UK. He said he was running regular courses in Seville, southern Spain, where the weather is good all year round, adding that the course normally takes a week.

Andy had come across as trustworthy. On the way back to London, I thought that if I was ever going to skydive, it would be with Andy as an instructor. Adrian had wanted to do this course for a long time. He said he would definitely sign up and urged me to join him—and bring along some friends if I wanted to.

I discussed this potential AFF adventure with two other friends, Damiaan and Leo. I had approached them because I knew they were always up for a challenge and were great fun to spend a week with. They both said that skydiving would scare the living daylights out of them, which I was glad to hear because it meant I was not that much of sissy after all! Leo wasn't in the best shape of his life but was pretty much game from the moment I launched the idea. Damiaan is very considerate and asked a lot of questions. I couldn't answer them all. "What do you do when your parachute doesn't open? Is there a reserve? What type of plane will we be jumping out of? How do we take the wind into account upon landing?" Although none of us really understood what we would be signing up for, we decided to go for it. I called Adrian, who was delighted with the news. This fearless foursome would soon be falling from the sky. We were going to realize our dream to fly!

We set off for Seville in April 1998, and the weather was beautiful as expected. The facilities on the drop zone looked impressive with a big landing area, a large packing room for the parachutes, and a number of Twin Otter planes that we would be jumping out of. The day we arrived, we were told we'd be doing our first jump in the afternoon; the morning would be dedicated to on-the-ground training. We knew that we would not be getting the six-hour training that one gets for static line parachuting, but we did expect more than an hour and a half!

We only went through the basics of the nylon rig or backpack that holds the parachute, which skydivers normally refer to as the canopy. We were told to handle the rig and canopy with great care so as not to damage it. We also learned that every rig also holds a reserve canopy in case of a malfunction. You would first have to disconnect the main parachute through a release system on your rig. Then you'd have to pull a handle on the harness of your rig, which would jettison your malfunctioning canopy and allow your reserve canopy to deploy freely. This action is called a cutaway. I sincerely hoped I would never ever have to do a cutaway.

Andy continued to elaborate on the basics of skydiving: "The rig also has an automatic activation device, which will save your life if you don't manage to pull your parachute yourself, for example if you pass out on your jump."

Well, hold on, I thought, *this all sounds way too dangerous.* But we were here in Spain now, I had taken a week off from work, and there really wasn't a way back anymore. But I wasn't comfortable at all.

Furthermore, we were taught by Andy how to steer the canopy with our toggles. And most importantly you have to make sure you get "the brakes" right: "You can brake by pulling both the right and the left toggle down at the same time. This is critical upon landing. Otherwise your landing will be similar to jumping from a three-storey building. And one more thing: make sure you land into the wind;

otherwise, your landing will feel like you're jumping out of a car at fifty kilometres an hour, and you'll break your legs." The last thing he advised us on was to ensure we had a stable arch position, with arms and legs stretched out, as soon as we jumped out of the plane to control our free fall. And that was about it. We then got to lie on a skateboard to practise the arch position while moving. I wouldn't say that at this point I felt totally ready for my first jump!

After a quick lunch, Leo and I were the first to take the Twin Otter to go skydiving. Andy welcomed us into the plane with a "Step into my office" and introduced me to the two instructors who would jump with me and hold me by the grips that I had on my skydive suit. They would help me to stabilize, and they would also give me in-air tips with their hands during my free fall. Andy explained that after my canopy opening at 1,500 metres, I should look at my altimeter and stop all manoeuvres by 500 metres. By 300 metres I would have to make sure to fly with the wind and reverse my direction 180 degrees. By the time you're at 100 metres from the ground, you have to land into the wind, so it is a landing pattern similar to that of an airplane. "I just want to make sure you don't break your legs," he kindly reminded me. I didn't know whether I would remember all that during my jump, but fortunately he gave me an earpiece that would allow him to at least give me directions for my landing.

And then the plane started to taxi to the tarmac. There really was no way back now. From here on everything went very fast. After take-off we started to gain altitude very quickly, and within no time I was hanging out of an open door at 5,000 metres. Andy shouted, "Are you ready to skydive?"

Not going to be honest, I shouted, "Yes!" Fortunately I was still thinking clearly enough to remember the procedure. I shouted, "Check in." Andy gave me the thumbs up. I then shouted, "Check out," after which it was "Up, down, arch." And the next thing I knew was that I was falling at a dazzling speed towards earth. Big plots of land looked like postage stamps. The feeling is difficult to describe, but think of yourself going down a roller coaster, multiply it by ten, and then you're halfway there. You feel it in your stomach, your cheeks blow up, you can hardly breathe, your throat gets incredibly dry, and the worst is that you keep accelerating! The wind sounds like a hurricane. The only comforting factor was that these two instructors had a firm hold on me. In about ten seconds I had reached terminal velocity and was falling at 180 kilometres per hour.

I remembered I had to show my instructors that I would be able to open my canopy before we got to 1,700 metres, which we had practised when we were lying on our skateboards: Look, reach, pull, arch. Somehow up here these practice drills didn't feel quite the same. As I saw the fields below me getting closer and closer, I couldn't remember any of what we had practised. The instructor pushed my hand towards the release handle, and it all started to come back. I touched it to let him know I could locate it on my rig. The next step for the instructors was to let go of me for a few seconds. I immediately understood the importance of arching, as I wasn't properly arched, which significantly altered my position. And you don't want to get into a spin because then it would be very difficult to stabilize yourself again. The reason it is so important is that you can't open your canopy when you're spinning because it will get tangled. Fortunately they grabbed hold of me again to put me in a stable position. After thirty seconds of free fall, my altimeter read 1,500 metres, which meant I had to release my canopy. For real this time! When I pulled the handle, there was a moment of panic. Nothing happened. What was going on? And then to my great relief I felt a big pull. When I looked up to double-check, I saw that my parachute had opened perfectly!

The next ten minutes were exactly the opposite of what I had just experienced during my free fall. It was absolutely silent—I couldn't hear anything—I could breathe properly, and my brain started to function again. I played around with the toggles, went through a

Checking the altimeter on my wrist on my first jump.

range of movements, and found that steering was actually quite easy. And fun! When I got to about 400 metres from the ground, I could hear Andy through my earpiece. He was going to help me land from where he was standing at the drop zone. He guided me perfectly. I was glad he was there to assist.

I quickly realized that you have to make sure you time your 180-degree turn into the wind well. If you turn too late, you might not be on time to land exactly into the wind. And if you turn too early, you might miss the drop zone. It takes practice, which obviously I hadn't had, so Andy's help was very welcome indeed. I also managed to flare quite well by pulling both toggles to slow my speed when I touched the ground. But somehow I still managed to land on the bloody tarmac instead of on the field, which was probably the size of a football pitch!

How on earth did I do that? I fell over and bruised myself, but to be honest I couldn't have cared less, as I had just experienced the ultimate rush: I had jumped out of an airplane! Leo had jumped after me but had already landed. We congratulated each other. We were all high-fiving. I asked Leo's instructors how it was possible that he had landed way before me. "We had difficulty keeping up with him during his free fall, as he was a lot heavier than he had told us before the jump. We will be using weight bags to keep up with Leo on his next jump!" Leo was standing next to us, and he wasn't too proud of this. I guess men get confused about their weight too!

For the second jump, Adrian, Damiaan, Leo, and I would all jump out of the same plane. We were starting to get a little bit more comfortable, Damiaan in particular. He would be the first one to jump. His instructor asked him whether he was ready to skydive, hoping for a positive answer, after which the normal routine would start again: check in, check out, up, down, and arch! Damiaan just looked at the instructor and said, "Are *you* ready to skydive?" And then he jumped straight out of the plane! His two instructors looked a bit baffled but immediately jumped after him. No more holding back for Damiaan; he was clearly getting the hang of it.

My exit went pretty smoothly. When I got into the arch position, my instructors released me, and I had to gently turn to the left and stay in control. And then I had to turn to the right. That was easier said than done, but I managed. The only thing I had to do after my turns was to get back to a stable arch position and maintain it for another ten seconds, with my instructors staying close in case I went into a spin. It was comforting to have them jump with me, but all went well. After I opened my canopy, I was actually quite relaxed and started to enjoy my floating through the air. I was absorbing the views and noticed again how quiet it was. It is odd, but you only know what silence sounds like once you've experienced it. It was almost as if time stood still, and it really allowed me to live in the moment. When those fields below started to look bigger, I had to remind myself that

I still had to get focused again for my landing, which fortunately went a lot better than the last time.

On my third jump I had to do more exercises from an arching position to really work on controlling my body position. It went all right, but I knew that the next couple of jumps would start to become more challenging. And I knew I would only have one instructor from now on instead of two.

On the fourth jump I had to do 90-degree turns from the arching position, which I did have difficulty with because I had no idea whether my turn had been 90, 180, or 360 degrees. After my first turn I had trouble getting back to a stable arch position. I started spinning when I tried to stabilize myself, which can happen very easily as the tiniest imbalance will have a great effect because you're falling at full speed. Fortunately my spinning was stopped by my instructor, who grabbed hold of me, after which I was able to safely open my canopy. My landing went well, especially when you take into account that I was no longer given a radio earpiece. But I knew that my jump had not been a good one given my spinning. I had to retake this level, which I was actually quite happy about because I realized I needed to practise my stabilization a bit more.

Once I passed this level, I got to level 5, where the challenge was to do a 360-degree turn to the right, stabilize again, do a 360-degree turn to the left, stabilize, and then release the canopy. I was glad that I managed to do these drills without my instructor's help, although he was still jumping alongside me.

During level 6 there would also be an instructor, but he would only help in case of emergency. Even during the exit I would be on my own, and the drills were even more challenging. Again I had to do 360-degree turns, but I now had to add a back loop, which I would have found challenging in a swimming pool, let alone during a free fall. I managed to do a barrel roll, but not more than that, and my 360-degree turn was actually 270 degrees. But with no sense of orientation, I thought I had done pretty well. Although my performance had not been impressive, my instructor deemed it sufficient to let me pass and move to the next level.

At level 7 I had to do exactly the same but now add tracking, which meant straightening my arms next to my body and extending my legs to become as aerodynamic as possible to get to maximum speed during the free fall. This position is referred to as "the Superman", and what an amazing feeling that was. Think of it: you're falling down to earth at maximum speed like a bullet, shooting through the clouds with your head down. It was just unbelievable. I liked it so much that I even did a second Superman, to the chagrin of my instructor, but I landed well with good canopy control. By now I was ready for my final jump. What could go wrong from here?

The plane was about to take off again, but there was no rigger available to refold my canopy for this last jump. A rigger is someone who is trained and certified to pack a parachute, and believe you me, you'd rather have a rigger fold your parachute than do it yourself. I said, "Andy, I'm sorry, but I'm not going to try to fold this canopy myself. I'd rather wait until a rigger becomes available."

Andy was quite resolute. "TJ, we have a good weather window now, and the weather is not looking great tomorrow or the day after. We've got to do it now." Andy gave me another rig that had already been prepared and told me we would be leaving again in ten minutes. This didn't make me very comfortable as Andy, only days earlier, had told us that one should always be present when the canopy one will take on the plane is folded. That is the only way to know for sure that there actually is a canopy in your backpack! But this was not the time to argue with Andy—that much was clear—so I took the rig and followed him to the plane for my graduation jump.

For this one we would go to an altitude of 5,500 metres to give me plenty of time to jump out of the plane, get into a stable arch position, and open my canopy. Andy got in the Twin Otter with me. He said, "There will be no instructor to jump with you, so good luck. You'll be fine. I'll see you at the drop zone!" I had felt relatively safe until that point because so far I had always had an instructor with me to help out in case it was needed. This felt completely different. And bloody frightening. I reminded myself that this jump was not about having fun, doing Supermans, or making exciting canopy turns. There would be no one to help me get stable or open my parachute for me and help me with my landing. *Just get down in one piece,* I told myself.

My exit was good, and I got into an arch position without any difficulty. I was not going to take any risks and wanted to open my canopy ASAP, so I pulled at 4 kilometres and was waiting for that comfortable feeling of suddenly hovering through the air. I was waiting and waiting, but nothing happened.

When I looked up, I saw that my canopy had actually opened, but the lines were twisted, which prevented the canopy from opening fully. Immediately I knew I had a major problem, but somehow I managed to become very calm and focused. I simply had two choices. Either I would manage to readjust the lines of my canopy, or I would have to do a dreaded cutaway and hope for the reserve parachute to deploy. The latter didn't sound too appealing. I hadn't been there when this rig was packed, so could I be certain there would actually be a reserve? I decided I'd better get my twisted canopy in order, and rapidly, as I saw the earth was now approaching fast. I was at 3 kilometres and would have to make a decision at 1,500 metres or preferably 2,000 metres. You can only open a parachute when you're falling at terminal velocity; otherwise it won't open properly. A cutaway at 2 kilometres would allow me to resume my free fall and reach terminal velocity again which would enable the reserve parachute to properly open—if there was one.

I was pulling hard on both toggles and couldn't figure out how the lines were twisted. They just were and didn't respond to any of my actions. I had a real problem here, and I needed to solve it in the next thirty seconds.

At 2,500 metres I managed to get one twist out by a hard pull on my left toggle, which gave me the encouragement to try again and again on both toggles. Finally, after another 400 metres of pulling and cursing, I felt the upward pull that I had been waiting for. My canopy had properly deployed!

This whole ordeal probably hadn't lasted more than a minute, but it felt like everything had slowed down and that this agony lasted for hours. I now started to get my breathing back and was very relieved that I had not been forced to cut away this parachute. To be honest, I'm not sure I would have had the guts.

No one at the drop zone had seen what had gone on up there in the skies, so when I landed, everyone was very excited. I had "graduated". Andy gave me a big hug and told me I was now officially a skydiver! He also mentioned one of his favourite quotes, which is from Leonardo da Vinci: "Once you have tasted flight, you will forever walk the earth with your eyes turned skyward, for there you have been, and there you will always long to return."

I felt proud at that moment and was very thankful that everything had worked out well, but at the same time I also realized I had gotten lucky. I will always agree that nothing gives you a greater adrenaline rush than skydiving. But it was also right there and then that I decided never to jump out of an airplane again. No disrespect to Leonardo da Vinci, but this was all the skydiving fun I ever wanted.

Adrian, me and Damiaan after our last
jump. We were skydivers at last!

Chapter 3

STILL AIMING HIGH

The tallest mountain in the Netherlands, my native country is, only 322 metres high, so mountaineering is not something I was born with. Nevertheless, I had always appreciated the beauty of the mountains when we went skiing or hiking when we grew up. But it wasn't until I was 26, when I was driving back from La Tania after a skiing holiday, that I passed the Mont Blanc, the highest mountain in the Alps, and was captivated. I asked our driver what it would take to climb it. He had to laugh as I had told him moments earlier that I had never climbed a real mountain before. His advice was not to even try, but he said that if I wanted more information, I could always call the Bureau des Guides. And so I did.

Not surprisingly I was firmly told by the Bureau that high mountains, like the Mont Blanc, are an unforgiving place and that this mountain should not be anyone's first high-altitude adventure. In fact, I learned that the Mont Blanc is one of the deadliest mountains in the world, and many climbers lose their lives trying to get to the summit. But at the end of our conversation, the guide I was talking to gave me some hope: "Normally we don't want any people without proper mountaineering experience on Mont Blanc, but we sometimes take novice climbers up, provided they have joined our week-long introduction course to get to grips with the basics of mountaineering."

Interesting, I thought. *These guides are commercial after all.*

"So what are the sorts of things that I'd learn if I followed this introduction course?" I asked.

"Well, you'll learn how to use an ice axe, a climbing harness, and crampons; what to do when you fall into a crevasse; how to avoid an avalanche or a serac [overhanging block of ice]; and how to deal with altitude. If you're fit, we could get you ready for the Mont Blanc in one week."

This all sounded pretty scary. The dangers of mountaineering were very clear to me. But at the same time I thought the summit would not be completely out of reach. After all, I was fit and would definitely undertake this adventure with a guide who would help me steer clear of any disasters.

Adrian was the first person I called to discuss this idea. I knew he was also looking for a new challenge now that we had received our AFF's in Spain and had run the New York City Marathon the year before. During our skydiving adventure his diabetes hadn't been a big issue, and I hadn't really thought about his diabetes when I talked to him about us taking on a climbing challenge. He immediately responded with great enthusiasm, so we were on for the Mont Blanc! Luckily he agreed that we should use a guide given our complete lack of climbing experience, after which we subscribed to the week-long mountaineering introduction course with the aim to conquer the Mont Blanc at the end. We were looking to become mountaineers this time!

I didn't really know how to train for an adventure like this, but I presumed a healthy level of cardio training wouldn't hurt. After a rigorous workout regimen, I was in good shape when I got to Chamonix in the summer of 1998, not really knowing what to expect.

That became pretty clear when we had our gear check. It was actually quite embarrassing. Before we set foot on the mountain, our guide wanted to carefully inspect our kit and clothing, so I showed him what I had brought. He said, "This is all well and good, but what are you going to wear on the mountain?"

"Well, I have this golfing jacket, plus a rugby shirt, a baseball cap, some ski gloves, and some camping pants." He shook his head in disbelief. Adrian didn't fare much better during his check. He had a typical British attitude to the cold—just ignore it—so he didn't have any proper climbing clothes either. Our guide then realized he was in for a long week.

Fortunately we could buy a fleece and rent a bit of kit, so a day later we had some protection against the cold, plus plastic mountaineering boots, crampons, and an ice axe, although we had no idea how to use those. We had a couple of other people on this course who were better prepared as they had brought warm clothing and some climbing gear. What was comforting for me is that they didn't have a lot of mountaineering experience either. We would all start from square one.

Before we even set off for the mountain, we had a meeting with our guide, who wanted to give us a bit of an idea what we could expect over the next seven days. "The summit of the Mont Blanc is at 4,807 metres, and altitude is something of which the effects shouldn't be underestimated. The higher you go, the thinner the air, which means your body has less oxygen to perform its tasks. The only way to deal with altitude is to let your body acclimatize, and you do that by climbing high and sleeping low to recover. During that process your body starts to produce more red blood cells, which enhances the blood's ability to transport oxygen through your body. This process will help avoid altitude sickness, which could result in nausea and headaches but could also develop into high-altitude pulmonary edema (HAPE) or high-altitude cerebral edema (HACE). Although HAPE and HACE are not that common in the Alps, one should take the dangers of these types of complications very seriously, as they can be lethal." That was something I had never even considered. Sure, I knew that the air gets thinner the higher you go, but I had not assumed that altitude itself could be that dangerous. But the guide wasn't finished.

"On Mont Blanc the biggest dangers for climbers are crevasse falls and overhanging seracs, but also hypothermia, frostbite, and avalanches. The annual death toll comes close to three figures. Too many novices attempt Mont Blanc, mainly because the mountain is so accessible from Chamonix. But they have no business being on the mountain."

It was clear that too many guiding companies are understating the dangers when they simply advertise a Mont Blanc climb as a long walk to the summit. Adrian and I were wondering at this stage whether a one-week course would really prepare us for the Mont Blanc. We couldn't deny we hadn't been warned at this stage!

We agreed we would still go ahead but take everything one step at a time. We could still pull out at the end of the course and not climb the Mont Blanc if we didn't feel confident enough. When we told our guide we were ready for our first session on the mountain, he started to laugh. "Before you even set foot on the mountain, you have to get familiar with the basics. The Mont Blanc is almost completely covered in snow and ice, so you need to know how to use crampons and an ice axe, and I don't think you do know how to use those." He was right. How do crampons work, those metal frames with spikes that you attach to the soles of your boots? How do you rope up to your teammates? How do you put a harness on so that it doesn't come undone during a crevasse fall? How do you tie a figure-eight knot to attach yourself to the climbing rope? We didn't have a clue.

To learn these basic skills, we would go by train from Chamonix to Montenvers, where one can find the largest glacier in France at an altitude of 1,913 metres. It is called the Mer de Glace (which means "Sea of Ice"), and it is 7 kilometres long and 200 metres deep. It can be accessed from the Montenvers train station by walking down a path that gives a flavour of the beauty but also of the inhospitable environment one can encounter in the Alps. What clearly stood out was the Aiguille du Dru, a very scary peak that is reminiscent of the shape of a needle. It looked pretty unclimbable to me but had been climbed many times before, our guide informed us. I really got the feeling I was out of my depth in this environment.

After a couple of hours of going through the basics of climbing, we were asked whether we'd like to know what it would be like to fall into a crevasse. *Not really,* I was thinking. We all had to be lowered down one by one into a crevasse and climb back out again. This was bloody nerve-racking. And cold. It was then that it became clear very quickly why crevasses are so dangerous; if you fall in (and hopefully the rope holds), you can get hypothermia and freeze within minutes, so you lose the ability to climb out again. The best thing would be not to fall into a crevasse altogether, which is easier said than done, as they are often covered by snow so you don't even know they're there.

After a full day, we had a good understanding of how to use our harnesses, our crampons, and an ice axe. The next day we started our acclimatization by climbing a number of smaller peaks in the area to allow us to put our newly acquired skills into practice. This was also a good way for us to get acquainted with the principle of mountain huts, which are very basic. Most of them don't have running water, but they do serve food and offer mattresses.

I really enjoyed this new type of adventure. The freedom of the mountains, the overwhelming beauty, the power of nature which emphasizes your own insignificance—it really made me feel alive.

These couple of days enabled us to prepare for our big climb at the end of the week, but what we didn't know was that our guides were also assessing us to divide us into two groups.

One half of us were going to climb Mont Blanc via the Goûter Route, and the other half were going to use the Trois Monts Route, the former being the easier route and the latter being more technical and dangerous. To be specific, the Trois Monts Route has many crevasses, and serac falls happen frequently. But the advantage is that it is more scenic as you will climb three summits to get to the top. Furthermore, you can start the climb from the Cosmiques Hut, which you can get to from the cable lift in less than an hour. So you won't have to waste precious energy the day before your summit attempt, as compared to the approximately five hours it takes to get to the Goûter Refuge from the cable car. Adrian and I had done well throughout the introduction course and were told we would be doing the more difficult Trois Monts Route. We didn't know whether we should be happy about that, as we didn't feel 100 per cent confident about our technical climbing skills, but we decided to take on the challenge anyway.

We started out the next day from Chamonix and took the Aiguille du Midi cable car, which gave us an altitude gain of 2,800 metres in only twenty minutes! This was a luxury that didn't exist in 1786 when the mountain was climbed for the first time by a doctor from Chamonix, Michel-Gabriel Paccard. It made us aware how much more difficult that first ascent of the Mont Blanc must have been. From the Aiguille du Midi station at 3,842 metres, we had to descend the east ridge of the mountain. This snowy crest is not more than 40 centimetres wide, but it is 60 metres long. What made it very daunting was that it is also steeply downward sloping. This was not what I had expected. The first day was supposed to be an easy walk to the hut, but this was different. If I were to fall off to the left I would end up back in Chamonix in no time, and if I were to fall off to the right, I would tumble down steep slopes towards the Vallée Blanche. This *arête* was breathtakingly exposed, and a fall on either side would be fatal.

We were told to put on our crampons so as to make sure we wouldn't slip. We would have to put into practice what we had learned: stand up straight and focus on every step with our feet pointing outwards. Furthermore we would all be roped up so if one person were to fall, the others could hold him. On this ridge there was nothing to hold on to though, so if one fell, it was reasonable to assume that the others would be pulled down too. I thought I'd better bring this up with the guide. I said, "Just before we start walking down, I just want to make sure I fully understand what I need to do if someone falls."

The guide was resolute: "If the person before you slips and falls off on the left, you simply jump off on the right. Anything else?"

Just to put this into context, we were still at the Aiguille du Midi station, we hadn't even started, and I was already anxious. What made it even worse was the fact that people were coming up as we were planning to go down. How on earth would we manage this? *Bloody hell*, I thought. If this is the sort of challenge we'd face just to get to the Cosmiques Hut, what would the actual climb be like?

Once we started our descent down this snowy pathway, I was able to find my feet quite easily, but it got pretty scary when we encountered those climbers going up. The crest didn't really allow two-way traffic; there was barely enough space to go one way. We managed to very slowly pass each other. No words were exchanged; everyone just focused on the job at hand. We were all relieved when we managed this rendezvous without accident. Once we got off the ridge, it was a straightforward forty-five-minute walk to the refuge. I was very glad to get there. But I kept thinking, *What else does this mountain have in store for us?*

After an early dinner, we went to bed around 8 p.m., but I didn't get a lot of sleep. The hut was cramped, the altitude was 3,613 metres, and we were lying next to each other like sardines. And we all were very anxious about the climb ahead. We got up not long after midnight and set off around 2 a.m. to make sure the snow would be cold and compact, which would lower the chance of us getting caught in an avalanche.

We were definitely not the first, as we already saw a whole line of climbers going up the mountain. Because it was pitch-black, everyone was wearing a head torch. I looked to the sky, which was absolutely beautiful with every star clearly in sight. When you live in London, you don't see any stars, so this already was an amazing experience. I was in awe when I saw this long string of lights going up the mountain. It felt like we were right among the stars!

It was also practical to have these climbers in front of us. We didn't have to do any route finding; we just followed those lights. Although our way through the Col du Midi was strenuous at times, it wasn't difficult from a technical point of view. This gave me a lot of comfort. I really enjoyed this first bit of the climb. But that joy was short-lived because not long afterwards we reached the shoulder below the Mont Blanc du Tacul, the first of the three summits.

We had been going for two hours now, and I couldn't believe what I saw: a sheer vertical ice wall 20 metres high. There was an alternative route, but too many climbers were queuing up, so waiting for them would take too long. Our guide made the decision for us to climb this ice wall, something that none of us had ever done before. He told us, "Trust your crampons. Simply front point and kick into the ice, and use your ice axe for balance." That sounded pretty straightforward, but would it really work? He would go first, he said. He'd find a hold for the rope (a belay) at the top of the wall and then lower his ice axe so that the next person could climb up with two ice axes. Adrian was at the front of the line, so it was up to him to show us how to climb an ice wall!

With those two ice axes and boots with crampons, Adrian did manage to slowly ascend. As he and I were roped up, I had to follow him up this wall when the rope was pulled tight. Although I had only one axe, I also managed to find enough balance and climb up—and I was kicking my feet as hard as I could into the ice. What worried me the most was that if Adrian were to fall, his crampons would jam into my head because I wasn't wearing a helmet. No one was. I just had a woolly hat. Also, if he were to slip, we both would fall, and it would very likely pull out the belay device which was screwed into the ice and was our only protection at this point. When I looked at our guide, who didn't have an ice axe anymore, he was standing on a narrow plateau and was struggling to balance himself. He wouldn't be able to hold us either if we fell. I said, "Adrian, you'd better not fall. You'll kill us both."

His reply was clear: "Well, TJ, because you're so slow, the rope is pulling tight, and that will make me fall, so don't be so bloody slow!" That was easy for him to say as he had two ice axes and I had only one, but I didn't have a choice. I tried to find any hold with my left hand so I could increase my climbing speed. This ice wall was scary stuff, and I was very relieved when we reached that little ledge, only to find out that we were just halfway up the wall and had to do the same exercise again! I was swearing. My left hand had gone numb from the cold, and Adrian was not comfortable either. My legs were shaking incessantly. Other climbers had given up and started to come down, which made it even more cumbersome for us to go up. We realized that if we wanted to get to the highest point in the Alps, we simply had to carry on, but this was not a fun experience at all. Once we got to the top of this wall, it already felt like we had climbed Mont Blanc. I was exhausted, and my legs were still shaking.

This had been stressful and tiring, but fortunately from here on the path seemed to slope gently downward for a bit. Nice for a change. We were now at the top of the Mont Blanc du Tacul at an altitude of 4,248 metres, which we started to feel. I had developed a slight headache, and I was very much out of breath at this altitude. Everything is just more difficult at high altitudes because you have less oxygen. I started feeling lethargic. But I was not the only one suffering from the altitude. Our guide had gotten stomach cramps and had to relieve himself. The ice fall had probably been stressful for him too, although he had made fun of me minutes earlier because of my disco legs! I was just hoping he would be OK again because I wouldn't know what to do without him. I certainly was not going down that ice wall on my own!

It was still dark and cold, but we could slowly see the sun coming up. This was a huge psychological boost and certainly gave us new energy. After twenty minutes, the path started to slope up again, and by now the sun had fully come up. This was a beautiful moment; the view was absolutely stunning. Because we were so high we were looking down on all those other Alpine mountains. The view was infinite, and importantly we were finally warming up. My left hand had started to thaw, which I was very pleased about. However, my new-found optimism only lasted until we reached our next obstacle.

We now had to climb up a 45-metre-long slope with an incline of 50 degrees, but fortunately there was a fixed rope in place that our guide tested. We were relieved that it was deemed strong enough to hold us. This climb was strenuous, but the fixed rope made our climbing a lot easier. I used my right hand to hold on to the rope, used my left hand for my ice axe, and kicked my crampons hard into the ice, which was starting to get a bit softer now that the sun was up. When we got to the top of Mont Maudit, the second summit, the exertion and altitude really started to hit us as were now at 4,465 metres. It would only be a couple of hours from here on, but nothing technical, we were told, so I started to relax. This was the first time that I assumed that we might actually make it to the top of Mont Blanc.

As we continued our climb, it became very apparent that Adrian was suffering badly. He had slowed down his pace and had to stop constantly. On the other hand, with the sun now shining and the technical part of the climb over, I started to get more energy, especially now that I knew we were getting closer to the summit. I was hoping it would be the same for him. We kept going, but the guide also sensed something was off. When we reached a rock formation called the Petits Rochers Rouges, Adrian stopped again to have a break. He then got his glucose monitor out of his pocket to measure his blood sugar. The guide looked very surprised. "Adrian, what the hell are you doing?"

Adrian tried to play it down, saying, "I won't be long. I just have to measure myself. I have diabetes. Just give me a minute."

Our guide went livid. "Why did you not tell me about this? You never mentioned anything, not even during your medical check-up. This is completely irresponsible. Blood glucose monitors don't work up here!" He was right; Adrian couldn't get the thing to work. These devices are only approved up to 3,000 metres, and we were well above that altitude.

We had a big issue now because Adrian didn't know how many carbs to take in or whether he should inject insulin. If we were to continue, we'd run the risk he would get a hypo, which would require an emergency evacuation, if such a thing were even possible. To leave him on his own and pick him up on our way down from the summit was not an option either. The safest choice would be to go down, but at this point we were so close, the summit was within reach. What to do from here?

I've always supported Adrian to push himself, but this time it was different. I understand why he hadn't told the guide about his condition, as he probably would not have been allowed on this climb, but his decision didn't impact only him; it also had an impact on his fellow climbers. I shared the guide's opinion that this was irresponsible. Adrian understood the situation we were in and decided to take the initiative. "Listen, I know my body, and I can do this. I might be slow, but I will make it, so let's not waste any time and get going."

The guide knew that this was not without risks. "I'll let you continue for now, but if at any point I get the impression you're getting worse, you will go down." Adrian realized that he'd better push himself as hard as he could, which worried me because I knew he could sometimes push a bit too far.

We took it slow, and the summit just didn't seem to get any closer. I was monitoring Adrian constantly, and I could see how much he was struggling. After another two hours, we finally did make it to the summit at 4,807 metres, and we were all incredibly proud and grateful to stand there. We had had our fair share of dangerous moments during our climb, and I couldn't feel more alive standing on the top of this mountain. I was very relieved that Adrian had not gotten into trouble during the last stretch of the climb, but then our guide reminded us that when you're at the summit, you're only halfway. And 90 per cent of all climbing accidents happen on the way down. Clearly we had to stay focused to get back down safely. We took some photos but didn't spend more than a quarter of an hour on the summit before we headed down.

It was warm, so I took my jacket off when we left the summit. I had drunk a lot of water to rehydrate, and I was buzzing with energy.

Feeling alive on top of the Mont Blanc.

The Alpine views from the summit were
the most beautiful I had ever seen.

We had, after all, climbed the highest mountain in the Alps. But Adrian seemed to be getting worse with every step. This surprised me as one normally starts to feel better as one goes down. When we got to the Col de la Brenva at 4,350 metres, Adrian had given it all and was struggling to move. Again he tried to measure his blood glucose, but to no avail. He could barely speak, and this started to worry me. He was now very close to getting a hypo. I knew from our experience during the New York City Marathon that he'd get to a point where it would be difficult for him to make rational decisions, which I discussed with the guide. "Adrian is getting worse. He can barely speak, let alone walk. He needs sugar, fast. What food or drinks do we have left?" Adrian was almost oblivious to what was going on. He just sat there on a rock, staring down.

The guide shook his head. "We need him off the mountain as soon as possible. Climbing down is no longer an option. We need an emergency evacuation." We were fortunate that the weather was clear enough for a helicopter to fly out and land on the mountain, but the bad news was that our guide didn't have any communication device. Now I got annoyed with the guide. How was it possible that he had forgotten to bring his walkie-talkie? The situation at hand didn't allow for lengthy discussions. I was told to stay with Adrian while the guide would descend rapidly to find a way to call for an emergency rescue.

It was early afternoon, so we still had quite a few hours before darkness would fall. But Adrian was not getting any better, and he really needed to get help quickly. After three hours we did see a helicopter approaching in the distance. I was very happy to see it indeed coming our way. When Adrian got into the helicopter, I knew he'd be safely back in Chamonix soon, where he could check his blood glucose and refuel.

This was a comforting thought, but at the same time this had been another close call—too close, in fact.

Staying focused during the descent.

That evening when we were having our celebratory dinner, I didn't get the feeling that Adrian took it too seriously what had transpired during our climb. He was just brushing it off. This didn't sit well with me. "Adrian, it was irresponsible and selfish not to have alerted anyone of your diabetes. Do you realize that you got very lucky that you were able to be airlifted off the mountain? If not, it could have been fatal for you!"

He didn't seem to grasp the seriousness of it all. "I'm glad indeed I got down safely, and I feel great again. Maybe we should start thinking about a new adventure. I've been looking into climbing up Kilimanjaro and paragliding down. What do you think?" Although Adrian and I had now run the New York City Marathon and become skydivers and now also mountaineers, I wasn't sure it would be wise for him to go up an even higher mountain. I suggested he should maybe reconsider taking such risks.

Despite the mishap, our climb had been an amazing experience. As much as I was convinced that I would never skydive again, I was equally certain that I would climb many more mountains as this had been a truly magical, humbling, and life-changing adventure for me. I had never felt more exhilarated than on the summit of Mont Blanc.

I was, however, reminded of the dangers of climbing Mont Blanc when I read in a newspaper article a month later that more than ten people had died in one week after heavy rainfall had left the Mont Blanc dangerously icy. Clearly, one has to tread carefully in the mountains. Nevertheless, standing on top of the Mont Blanc was one of the most gratifying moments of my life, and I wanted many more of those!

Chapter 4

PEAK PERFORMANCE

After my first real mountaineering experience, I couldn't resist the call of the mountains. In the following years I went back to the Alps as often as I could, but I had started a new and demanding job which left me with little free time. Slowly I started to improve my climbing skills. I did away with the golfing jacket and began to learn more about the mountains whenever I found the time. Always keep the climbing rope tight as you want to immediately hold your fellow climber in case of a fall. Do not lean back when walking down, and of course, don't eat yellow snow.

Now that I was getting more confident on snow and ice, I still had to improve my rock-climbing skills. This became painfully clear when I was climbing the Dent du Géant, which is in the Mont Blanc massif. The summit has the shape of a tooth, and you have to climb the steep Burgener slabs to get to the summit. I knew I shouldn't use my arms; I should use my legs instead because they are much stronger. However, I couldn't get up the last section because I had difficulty placing my big mountaineering boots on the tiny footholds, and the fixed rope was so thick that I could hardly grab hold of it. As a result I didn't make it to the summit. I realized that if I wanted to take my mountaineering to the next level, I'd better work on my rock-climbing technique. I wouldn't be able to do that in the Alps because I lived in London with my new, time-consuming job, so I started training on weekends in a climbing centre called Westway. This certainly improved my agility and confidence, which is what got me started to think about my next climbing challenge at the end of 2008.

The Mont Blanc, the Matterhorn, and the Eiger are known as the big three of the Alps. And they are on every climber's wish list, if the climber is serious about his climbing. But none of these should be underestimated. The Mont Blanc is the easiest of the three and the Eiger the most difficult. But the step up from the Mont Blanc to the Matterhorn is already a very significant one. I wouldn't ever consider the Eiger, but would I be able to climb the Matterhorn? I started to do some research on the Matterhorn and read that this mountain has claimed the lives of more than five hundred people. But I was captivated at the same time because there is no other mountain in the world that is as iconic as the Matterhorn with its pyramidal shape. If one were ever to attempt to climb this giant peak, one would need to be in the best shape of one's life and deliver a very strong performance on the mountain to summit and make it back safely the same day.

The Matterhorn, the most iconic peak in the world.

This mountain was only conquered in 1865 by an Englishman called Edward Whymper, about eighty years after the Mont Blanc was first climbed. Whereas the Mont Blanc's first ascent started the golden age of alpinism, the Matterhorn's ascent was the last of that era. The reason it took so long for anyone to successfully scale the mountain is obvious. Simply take a look at any picture of the Matterhorn and you can see how steep and scary this mountain is.

Edward Whymper's first ascent is both famous and notorious. Famous as the Matterhorn was the last great mountain in the Alps to be climbed despite many attempts. And it is notorious for the tragic events that happened during his descent. The first endeavours to scale the Matterhorn had taken place on the south-west ridge, but when Whymper tried to climb from the north-east side along the Hörnli Ridge, he knew he had found the right route. On his many attempts he got closer and closer, but it wasn't until 14 July 1865 that he and his six fellow climbers would reach the summit. When they looked down from the summit onto the Lion Ridge, between the south face and the west face of the mountain, they saw an Italian group also making their way. Elated to have the first ascent to his name, Whymper claimed victory by shouting abuse at the Italian team, and in an unsportsmanlike way he kicked a rock their way, which wasn't well-received.

Whymper's glorious victory was marred by what was about to unfold. When the team started their descent, they were exhausted. On their way down a young novice climber by the name of Douglas Hadow slipped. He knocked over the French guide Michel Croz, and the weight of these two men pulled down Charles Hudson and Lord Francis Douglas as the entire climbing party was roped up. At this stage the whole group was about to be pulled off the mountain, but then the rope snapped. Whymper and two climbing partners were able to cling on to the mountain as they watched in horror how their four fellow climbers fell to their deaths.

This led to a huge debate in climbing circles, and many accusations of excessive risk-taking were levelled at Whymper. But the bigger question that was now being asked related to the justification of climbing high mountains; is death a price worth paying to climb any mountain? The press was relentless, but at the same time the people were captivated by mountaineering. One thing was certain: the ascent of the Matterhorn established its infamous reputation.

A more recent example of that debate was the 1996 Everest disaster that got worldwide media attention, especially after the publication of Jon Krakauer's book *Into Thin Air*. That year a number of high-profile climbers were all trying to summit, and pressure was on the guides to make that happen. Due to a lack of fixed ropes on the mountain, those climbers who were looking to summit on 10 May were slow in their ascent. At the same time, unbeknownst to them, a storm was brewing. Many got caught in a blizzard and were unable to go down, which led to the tragic death of eight climbers. Guide Rob Hall was one of them. He had refused to leave his collapsed client Doug Hansen behind, although it would not have been possible to save him at this altitude under these circumstances. The two of them were well above 8,000 metres when they got caught in the storm. They were not expected to survive the night, but somehow Rob Hall did. The next morning he could barely move and his hands were frostbitten, but on his radio he managed to get through to base camp, who were able to establish contact with his wife in New Zealand, who was seven months pregnant. He knew it would be his last conversation, but he reassured her that he was comfortable. He said, "Sleep well, my sweetheart. Please don't worry too much." He perished moments afterwards. The death toll would have been nine if it hadn't been for Beck Weathers' fierce determination to survive. He was found on three different occasions during the rescue mission but was considered beyond help every time. Miraculously he was able to get up again and again after having been left for dead. Ultimately he was flown out from Camp 1. He survived but lost the five fingers

Many Worlds to Conquer

on his left hand and half of his right arm to frostbite. When disasters like 1996 on Everest happen, the same issue always comes up: is climbing mountains really worth it?

Despite that ever-lingering question, I decided to go for the Matterhorn in early 2009 and called Olly Allen, with whom I had been climbing a lot over the previous years. He's one of the best guides in the Alps. I asked him whether he'd be willing to guide me to the summit of the Matterhorn. He knew my climbing level and said I would probably be able to do it but added that I would have to spend a week before climbing other peaks for optimal preparation and acclimatization.

I started working hard on my fitness, not just in the gym but also at the Altitude Centre in London, where I was training in a hypoxic chamber that simulates being at altitude as the air in there is thinner. As in the mountains, your body will start to produce more red blood cells, which carry oxygen, which will lead to increased performance at altitude. This would give me a big advantage in that I would already be acclimatized once I reach the Alps in early August, which was when I planned this next adventure. By midsummer I felt physically and mentally ready for the challenge ahead as my training on the Westway climbing wall had also boosted my confidence. And I had stopped using elevators a month before the climb; I only took the stairs.

When I got to Zermatt, I was very keen to get a glimpse of the mountain. I had never been to Zermatt, let alone seen the Matterhorn. When I got out of the train station, the mountain presented itself immediately. And what a sight it was! It is high, steep, and incredibly daunting. I asked myself whether this was really such a good idea, but at this point there was no turning back. In the evening I met Matt Dickensen, whom Olly had arranged as my guide given he already had prior commitments that week.

Matt is a tall and strong guy from Derbyshire. We immediately hit it off. Although his climbing résumé was more than impressive, he was very humble and unassuming. He didn't need to show off. I liked that. We went for a burger, and I said to him, "I presume you have climbed the Matterhorn before. How difficult is it?"

He said, "It all depends on the route. If the weather permits it, we'll be climbing the Hörnli Ridge, which you should be able to do. But if you'd like to do a different route, just let me know."

Well, I wasn't really looking for a more challenging route than absolutely necessary, but I asked him anyway what alternatives there were. "Well, we could consider the north face if you really wanted a challenge."

I started to laugh and said I would never even think about it. But then I asked, "Have you actually done that?"

He took another bite from his burger and said, "Yeah, I did that solo, unroped, and in winter. Quite good fun." Matt clearly was the real thing. I couldn't wait to start my Matterhorn adventure with him.

After we finished our food, he explained we would climb Castor at 4,228 metres and Pollux at 4,092 metres first. Both these four-thousanders would help me further acclimatize. They would also be a warm-up for the two mountains after that: a traverse of the Breithorn and the Riffelhorn, which is also known as the Little Matterhorn because the climbing terrain is pretty similar as on the majestic peak. Only then would we attempt to climb the Matterhorn.

During our first two climbs I already realized that I had signed up for some serious mountaineering. As the week progressed, the terrain got steeper and the climbing more technical. The beautiful thing about climbing is that it forces you to get out of your comfort zone. When

Climbing Castor allowed me to further acclimatize.

you're stuck on a ridge, you can't just give up and take the bus home. You have to climb your way out. On these ascents I had to climb up steep ice walls, descend vertical faces, and work on my surefootedness. My acclimatization at the Altitude Centre was paying off as I wasn't getting too tired, not even at 4,000 metres. I was surprised to see that Matt was actually struggling. On our way down from the Riffelhorn, he said I was ready for the Matterhorn and then continued: "I'm not sure I will be able to guide you. I'm feeling weak. I probably caught the flu."

I knew how strong he was, and I felt comfortable climbing with him. "Even if you're not 100 per cent, you could still make it. Are you sure you can't guide me?" I asked.

"TJ, you're about to take on a very serious mountain. I don't want to jeopardize your safety. I will try to arrange another guide and check who's available the moment we get back to Zermatt." I very much appreciated his honesty, but I hoped I'd be able to climb as fluidly with another guide as I had climbed with Matt. You're constantly roped up and have to climb as a team if you don't want to fall off the mountain.

The next day, 13 August, Matt introduced me to Tomaz Jakofcic, who is a renowned Slovenian climber. He's incredibly strong, looks like Brad Pitt, and is also very sympathetic. I found out that his wife was about to give birth back home in Slovenia, so Matt had clearly called in a favour to have Tomaz guide one more climb that season. I was very honoured to have him as my guide and thanked him in advance for putting up with all the headaches I would undoubtedly give him.

We took the Gornergrat train from Zermatt around lunchtime that day to start my intended Matterhorn climb. On our way up I got the most beautiful view of this peak. The closer you get, the more you realize how incredibly big this mountain is. I was in awe, but at the same time I started to feel anxious of what lay ahead of me. We got out at the Schwarzsee station, from which it took us a couple of hours walking to get to the Hörnli Hut at 3,260 metres. From here we would start our summit bid the next day. As dinner would not start for another two hours, I made my way to the steep wall where our climb would begin. It's only a five-minute walk from the hut, and you can't miss it as there are fixed ropes in place to help you. It did not look like I was going to have an easy start the next day; this was a short but steep slab. I tried to climb up a couple of metres and actually did OK, but this was in clear daylight and there was no one in front of me or behind me. I knew it would be a different story the next day.

We were looking to start early to give ourselves as much time as possible to get to the summit and get back down again in daylight. After our 3.30 a.m. wake-up call, followed by a quick breakfast, we were ready to go at 4 a.m. But so were all the other climbers! I found out there is a strict hierarchy: Swiss guides and their clients go first, and then everybody else. Tomaz had just climbed Everest and had clearly earned the respect of the Swiss guides, who all wanted to have a chat. That's how we got to the front of the "foreigners'" queue and were able to leave the Hörnli Hut at 4.15 a.m. When we got to the fixed ropes, I could see the anxiety among the climbers as everyone looked clumsy climbing this bit. Once I got up this slab, things got a bit easier. We were on a ridge and traversed left to the east face, after which we climbed a couple of couloirs. After an hour of scrambling, the sun started to come up, which lifted my spirits. We were mainly climbing rock, rather than snow and ice, so the sun wouldn't cause dangerous slushy snow on this mountain. I was feeling pretty comfortable.

However, I had been advised by Matt before I'd set off with Tomaz to climb as fast as I could for the first couple of hours because most guides will turn you around if you haven't reached the Solvay Hut, the halfway point, within three hours. If you're not fast enough, you will not be able to make it to the summit and back before nightfall, and that is when things can get very dangerous, climbing down a steep ridge in the dark. Fortunately I was well acclimatized and felt strong through the first hour, which Tomaz was pleased to see. Nevertheless, he wasn't going to stop for a break until we reached the Solvay Hut.

Just before we got there, we encountered a very steep rock wall known as the Moseley slabs, where I was happy to see fixed ropes in place. As these ropes were very thick, it was not easy to get a good hold on them, which is why a little traffic jam had formed here. It was then that I realized Tomaz and I had been climbing alone until now. Somehow everyone loses each other on the mountain, only to see each other again at the bottlenecks. I didn't mind having a little wait here because I was able to catch my breath again before it was my turn to go up. Fortunately I was able to place my legs well when it was my turn, so I merely used the ropes to balance myself instead of pulling myself up. My training over the past week had paid off. I was able to climb this wall in style!

From the top of this cliff to the Solvay Hut was a short climb. When we passed the Solvay Hut at 4,003 metres, I noticed it is a very basic emergency shelter which was not designed for overnight stays like the Hörnli Hut. Tomaz told me that sometimes climbers do actually sleep there to shorten the climb on summit day, obviously something guides disapprove of, as it should only be used for emergencies.

The moment we left the Solvay Hut and continued our summit bid, I felt confident because we were now closer to the summit than to the bottom of the mountain. I didn't think of the summit yet because I was too focused on being sure-footed, especially when we reached the scariest part of the climb: a steep exposed ridge that was as narrow as my boot, with a 1,000-metre vertical cliff on the left and a flat rock face on the right with nothing to hold on to or use to anchor yourself.

The most dangerous aspect of the traverse across this ridge was that Tomaz and I were not secured. On the Matterhorn you're constantly moving together on a short rope, but there are only a few places where you can anchor yourself to a rock or something solid. And when you can do this, it will slow you down significantly. As a result you don't really anchor yourself, so there are not many moments when you're safe on the mountain, and definitely not on this narrow ridge with sheer exposure. I weigh over 200 pounds, and if I were to fall, I would likely pull Tomaz with me. This was the first time I saw Tomaz look anxious, and I thought I'd better focus. After a very intense crossing, I'd made it safely to the other side, to my and Tomaz's great relief!

He assured me we had one more tricky bit ahead of us, after which the summit would be in reach. I was glad he hadn't told me before of all the difficult sections because I would have had too much to worry about. Sometimes ignorance is bliss!

I still felt strong and again realized that my body was well acclimatized. After more scrambling, we found ourselves at the bottom of a large shoulder on the ridge leading up to the summit. We saw that many climbers were going up this steep rock face very slowly despite the fixed ropes that were in place on the most-exposed sections. They had given it all in the first few hours and were now paying the price.

Tomaz looked at this patiently, but after three minutes he decided we had to climb up and overtake them. The unavoidable passing manoeuvres on this 100-metre-steep section are not without risks, as you have to let go of the fixed rope to climb past another climber, and sometimes the others will push you away as everyone is jostling for position. We were climbing up a wall of rock and snow, and here every climber was wearing crampons. The last thing I wanted was someone falling on me or kicking me with their metal spikes!

I was pleased that my energy levels were still high, and I was wondering how some of the others would get to the summit and make it back safely again. It was only at this point that I dared to think about actually making it to the summit of the Matterhorn. We were getting really close now! When we got to the end of these ropes, we didn't have to climb much farther before we reached an ice field at about 200 metres from the summit where a St Bernard statue is placed. I wanted to take a picture, which Tomaz didn't allow because we needed to get to the summit as fast as we could. Only there he would allow a quick photo, after which we would need to descend quickly.

At exactly 9 a.m. on 14 August 2009, we stepped onto the summit of this iconic peak. Tomaz seemed as happy as I was. Guiding on the Matterhorn is no mean feat, especially with an amateur climber roped to you! It was a beautiful day with clear blue skies, and I could see the Mont Blanc, where my love of climbing had begun. At only 100 metres away I saw the Italian summit marked by a big cross. Some climbers were just reaching that summit as I was standing on the Swiss summit. There is a steep, rocky, and snowy ridge that separates the two. I was glad when Tomaz explained that this Swiss summit was the highest of the two so I wouldn't have to make my way over. Thank God for that.

On the summit of the Matterhorn with Tomaz who is still feeling strong! You can see the Italian summit on the right.

When I climbed the Mont Blanc more than ten years earlier, I never would have even dreamed to be standing on top of the Matterhorn. I was very thankful to have climbed this beautiful peak. We were alone on the summit, and I tried to take it all in for as long as I could, but we didn't have time to waste. We still had to get down, which on the Matterhorn, some people say, is harder than getting up!

When we were about to start our descent, we met two German climbers who were now setting foot on the summit but had forgotten their camera. They asked whether I would take a picture of them, which I did, but Tomaz told me to go down immediately, so we never got to exchange contact details. I still have that photo, and I know how valuable it would be to them. If it's you reading this, please let me know and I'll send you the picture!

I never got the details of the German
climbers whose picture I took on the summit.
If it's you, please let me know!

The first part of the down climb wasn't difficult. When we got to the fixed ropes on the shoulder again, Tomaz wanted me to go down next to the fixed ropes so we wouldn't have to deal with those climbers who were still going up. The technique he had in mind was an abseil, which is a controlled descent from a cliff while hanging back on a rope that is attached to an anchor at the top. You simply

Practising my abseiling down the Riffelhorn.

hang back and place your feet for balance. You don't need to use your hands; you just have to fully trust the rope to hold you. And that can be daunting going down a vertical cliff! Fortunately I had practised abseiling enough on the Riffelhorn to be comfortable.

After a few abseils, we decided to keep descending this way because this was much faster than climbing down. We were making good progress. Tomaz started to increase the speed at which he was lowering me. I didn't have a problem with that, but I wasn't able to control the speed of my descent other than to shout at him. And this became an issue when we came to an overhanging cliff. When you abseil an overhanging wall, the goal is to land a rope length down on a shelf or plateau. If there is one.

Neither of us knew what I would encounter the moment I went over that overhanging edge. I told him to lower me slowly. I carefully went over it but didn't like what I saw: there was no plateau, and this overhanging wall was also angular, which meant that gravity would prevent me from going down straight. I would be pulled to the left every time I would use my feet to push off the mountain.

My fears were confirmed when I tried to balance myself going down. I was immediately pulled sideways. Tomaz, who was standing on the top of the cliff, didn't see that, and he started to lower me faster and faster. Not only was I picking up speed by going down, but also I was picking up speed from swinging right to left. I shouted at Tomaz to slow it down, but he couldn't hear me. I was now gaining momentum so rapidly that it became impossible for me to control my position. I had started to swing violently while descending at the same time and was now effectively dragged along the wall. After 20 metres I started to lose control and realized I was in real trouble.

Tomaz still didn't know what was going on, and there was no way I could stop myself. The only way to come to a halt was to crash against a rock. The intensity of the moment was such that it gave me a level of concentration that I had only had once before, during my skydiving adventure when my parachute didn't open. Just like then, my thoughts became very clear, and everything started to happen in slow motion. I was aware of the situation I was in, of every breath I took, and I started to anticipate what would happen next. There was no other outcome than a forceful smash into something. Next thing I knew, my head hit a big rock, after which I landed on a ridge.

Only now did Tomaz realize that something was wrong below as the tension on the rope had gone. He immediately looked over the edge and got very worried when he saw me lying there. At least I was lying on a plateau with the rope attached to my harness. Much better than seeing a cut rope with a climber falling down the mountain. What worried him was that I didn't move, as I was dazed after my smack. But then I slowly opened my eyes. I felt my arms and legs and realized nothing was broken.

When I got up, I didn't hold back. "Tomaz, you should have controlled the rope. You almost killed me. Are you fucking insane?" At this point Tomaz realized I wasn't totally incapacitated. I've rarely seen anyone happier being shouted at! I myself was relieved to be alive, but all the euphoria from the summit had gone. I was shaken, and I realized when I started to move that I had torn my left thigh and could barely use my leg. We still had to get down.

From here on we would no longer abseil and would simply climb down. I didn't mind that as I had been told before my climb that abseiling is a bad idea anyway on the Matterhorn. But Tomaz and I had been overconfident, which is dangerous, especially on the Matterhorn. I learned the hard way that this mountain is a great leveller, and this incident reminded me of the fact that the Matterhorn is one of the deadliest mountains in the world.

The fastest way to climb down rocks is to face downwards, but it's also very scary. Your intuition tells you to climb down in the same position as you climbed up so you'll have two footholds and two handholds. But it's slow on the way down because you can't see below you. We needed to get off this mountain and had a long way to go, so Tomaz said I would have to face downwards and get on with it.

I was in terrible pain and tried to go as fast as I could, but with every step I took, my thigh and my groin caused me a tremendous amount of pain. Tomaz was tough on me and wouldn't let me slow down because he knew we had to be down the mountain before darkness fell. The next seven hours were an absolute agony. I was hurting badly and had to concentrate on every step I took. We finally made it to the bottom of the mountain at 5.15 p.m., and it was only then that I started to enjoy the fact that I had actually climbed the Matterhorn.

When I checked my helmet the next day, I saw there was a very big dent where my temple was. The impact of my fall would have probably killed me had I not worn a helmet, like on the Mont Blanc, and I was very thankful to have made it down safely. Back in Zermatt, Tomaz had to leave for Slovenia straightaway because his wife was going to give birth any day now. I thanked him and reminded him that I had warned him before our climb that I would give him headaches.

After our farewell I called Matt, who was as excited as I was that I had successfully climbed the Matterhorn. He told me, "I'm so sorry not to have been there with you at the summit, but I'm so happy that Tomaz got you to the top in one piece. And back! Do you realize that you've just accomplished what most mountaineers would see as the pinnacle of their climbing career? You just showed up and did it!"

I thanked him for his kind words and explained, "Matt, whatever Tomaz told you, this wasn't plain sailing. I had a bad fall on the way down and just made it down before darkness. I'm limping as we speak." He told me I should focus on the good rather than the bad and told me it would only sink in at a later stage that I had actually climbed the Matterhorn. He was right.

Back in London I started to reflect on this climb that had taken me out of my comfort zone more than any mountain before had done. I had come very close to a fatal accident, and now I asked myself, *Was it really worth it?* But as time passed, my memories of the Matterhorn were more about that tremendous feeling of standing on the summit than about my fall on the way down. And then I started to think. *Would I be able to push myself even further and climb a mountain even more challenging, like the Eiger, one day?* There's a saying in mountaineering: the best climbers are the ones with the shortest memories.

Chapter 5

COMPLETING THE TRILOGY

It was July 2011. I had just come back from Denali, the highest mountain in North America, but I was utterly disappointed. I had spent three weeks on this mountain in Alaska but hadn't summited because bad weather had forced us to retreat. Such is life, but it was a

Down from Denali which I didn't summit.

very big setback. I had trained very hard for this climb. The one good thing was that I was now in the best shape of my life. On the mountain I had lost 9 kilograms. My body fat percentage was 10.2, and I wanted to use my fitness for another adventure. I knew that if I ever wanted to climb the Eiger, I would need to be in the shape I was in, so I started to entertain the idea.

The first time I saw the Eiger was in 2005 when I was climbing the Mönch and the Jungfrau in the Bernese Oberland. From the terrace of the Kleine Scheidegg Hotel, I had gazed at the north face of the Eiger, and it looked dark, unwelcome, and scary. I couldn't imagine that anyone would be foolish enough to climb it. The Eiger is one of the most written about peaks in the world, not in the least for all the tragic events that have happened there over the years. Interestingly, the first ascent, which took place in 1858, went pretty much unnoticed. An Irishman called Charles Barrington had climbed the west flank, but it was only thirty years later that his achievement was actually published in the *Alpine Journal*.

It took more than seventy years after that first ascent before the Eiger started to attract real attention. It definitely wasn't because of its height, as the Eiger is 3,970 metres and fails to reach the magical 4,000 metres. It was because people started to attempt to climb the north face in the 1930s. Every party failed, and Edward Strutt, the president of the British Alpine Club, stated that climbing the north face of the Eiger was for the "mentally deranged". In the space of three years, no fewer than eight climbers died during their attempts, succumbing to cold, exhaustion, and falls. After that, climbers and the wider public started to become fascinated with this mountain.

The most tragic story of all happened in 1936 when four climbers had to retreat from the north face, which was impossible without a rope in place. Three died on the descent when they were caught by an avalanche. Most tragically, one climber, Toni Kurz, was close to being rescued but sadly his rescuers couldn't get to him. He was hanging from the rope and was in shouting distance from the rescue party, only to find out that he couldn't lower himself anymore because a knot in the rope prevented him from abseiling any further. He had to spend a whole night hanging on the rope, slowly freezing to death. When his rescuers returned the next morning, he was still alive but wasn't able to move anymore as his hands and body were frozen. Just before they got to him, they heard him say, "I cannot go on." And then he died, moments away from being saved.

The north face of the Eiger was finally climbed in 1938, which was immortalized in a book called *The White Spider*. When the Austrian Heinrich Harrer and three fellow climbers made that first ascent of the north face that year, the Nazis incorporated the success on the "Eigerwand" into their propaganda. The four climbers met Adolf Hitler and even took part in the regime's celebrations. The story of the controversial Heinrich Harrer was the basis of the film *Seven Years in Tibet*.

The mountain was also the subject of another movie in 1975 that was called *The Eiger Sanction* with Clint Eastwood in the lead, which certainly helped to put the Eiger on the map with climbers and nonclimbers alike. In short, the Eiger, and in particular the north face, has been captivating the public for almost 100 years, as it is loaded with lore.

I was very much captivated too by the history of this mountain, but I also knew that the north face was not something I would ever consider. It's simply too dangerous. Fortunately the Mittellegi Route and the southwest ridge offer good alternatives if you want to climb this mountain. The other alternative route, the west flank, which was used on the first ascent, is rarely climbed anymore, mainly because of hazardous rockfalls.

Having just returned from Denali, I found that everyone in the office was asking me about my climb. It was very frustrating to tell my colleagues I hadn't made it to the summit. I felt low and looked up how many holidays I had left and realized that I could squeeze in one more short trip, at the end of August when we had a bank holiday. I decided to go for it and attempt the Eiger, but I wasn't going to tell any of my friends or colleagues. I called Olly, my climbing guide buddy, to see whether he'd be willing to take me to the summit. As expected, he was booked up the whole summer, but he arranged another guide for me, a Frenchman by the name of Rémy Lécluse. I was on!

This really cheered me up because I now had another adventure to look forward to. At the same time I realized that the chances of success were very slim. This mountain is very unpredictable, and conditions need to be absolutely perfect as you're constantly climbing on a very exposed and narrow ridge on both the Mittellegi and the south-west ridge. Too much snow or strong winds can make this mountain very perilous.

I still had about six weeks and trained hard to stay in the shape I was in. I did a lot of upper-body work in the gym and also used the Altitude Centre again because I knew I wouldn't have a lot of time to acclimatize for this adventure. Again I stopped taking the elevator and used the stairs wherever I could.

I arrived in Chamonix on Wednesday evening, 24 August, and Rémy picked me up on Thursday morning from my B & B for a warm-up climb. He was a couple of years older than me and looked tough as nails but had a very gentle demeanour. He was also very modest, which is often a trait of the most-accomplished guides. I told him I lived in London, and we immediately hit it off as he was an Anglophile.

He took me to the Arête des Cosmiques, which I had climbed years before with Olly. I didn't feel great because I had picked up some bug in London, but I managed well, and Rémy was pleased to see I could hold my own on the mixed terrain of this ridge. The next day we went to Montenvers, where we were going to practise steep rock climbing, and I was glad I had brought my rock-climbing shoes, which made it much easier than trying to climb these near-vertical walls in big climbing boots. I still wasn't feeling great, but I also managed these slabs quite well, and Rémy said I would be ready to take on the Eiger.

Climbing the steep walls of Montenvers.

He was an incredibly gifted climber, but his real passion was extreme skiing, he told me. He would casually draw my attention to the slopes he had skied down from, and I just couldn't believe anyone would be able to do that, ski down almost vertical slopes. He pointed to many mountain faces that had never been skied by anyone but him, and he also explained that on those type of descents you can't afford to fall. "If you do, it's fatal." I thought he was absolutely nuts, but I was intrigued at the same time. What a guy!

When we got back to Chamonix that evening, I said, "I don't think we should get ahead of ourselves, but suppose we do make it to the summit of the Eiger and back in one piece, we should definitely celebrate it. You're French, so I presume you like wine." He confirmed he had a passion for fine wines. I continued, "You tell me what your favourite bottle is, and we'll have it."

He started to smile and didn't have to think for a second. "Cheval Blanc," he said. I had never had it before, but I had heard of this Bordeaux as it is often mentioned as the best red wine in the world with the very prestigious rank Premier Grand Cru Classé A.

"That sounds great. I would love to try it once," I said.

Rémy replied, "I've had it only once myself, but I'd love to drink it again. Let's make it happen!" We then went to bed because we would have an early start on Saturday.

That next morning I was glad to feel strong again and no longer suffered from the bug. The mountain air had done me some good, and I felt ready for the Eiger. We drove to Lauterbrunnen and made our way to the Kleine Scheidegg station, from where we took the Jungfrau Railway. After a short hike we arrived at 3 p.m. at the Mönchsjoch Hut, which is one of the more luxurious huts in the Alps. We relaxed a bit and had an early but big dinner, definitely no alcohol yet. I reassured Rémy we would more than compensate for that upon our return. We'd have a very early start again on Sunday and didn't make it late.

Over the preceding two weeks no one had been able to climb the Eiger due to bad weather, but things had started to turn in our favour. The weather had been getting better and better over the last couple of days. When we woke up on Sunday, I couldn't believe my luck. We were going to have perfect climbing conditions throughout the day: no winds, a clear sky, and no fresh snow that would make the ridge slippery. There were twelve other climbers in the hut who were all as excited as I was about the window we had been given. We all rushed to get ready.

Rémy and I didn't set out for the south-west ridge until 5 a.m., and it soon became apparent why he hadn't wanted to leave the hut earlier. We first crossed crevassed snowfields and then the Fiescher Glacier, all in darkness. Around 7 a.m. we got to the actual ridge, which allowed us to start on this exposed crest exactly when the sun came up, which gave us good visibility. "All those people who left before us must have struggled on the ridge in darkness. It's incredibly dangerous as you can't see a thing," Rémy explained. That had certainly been sound judgement as the rocky ridge that was ahead of us looked very daunting, even with full visibility.

I knew I was in for a long climb, with the ascent normally taking as long as the descent. If we'd climb fast we would make it back before darkness fell, but if we didn't climb fast enough, Rémy would make us turn around. He made that very clear, so I knew I would have to push myself really hard and climb with speed if I wanted to make it to the top of the Eiger. This also explained why the others had gone out before us, but I was happy that we had been responsible, even if it meant that I would really need to be speedy on the way up and down.

The beginning of the climb along this ridge wasn't too technical and was mostly on hard limestone rock, which gave me good footholds and handholds. It was very exposed though, so concentration was of utmost importance in order not to fall. My scrambling went well. I felt strong, and we started to overtake some of the other climbers who had started before us. After an hour of climbing, the ridge was getting more technical and I still managed well, but I saw a couple of others who were struggling. In particular we had to pass one young English climber who wasn't comfortable at all after he had fallen moments earlier, only to be saved by the rope that his guide had pulled tight. He was so shaken that he could barely move anymore, so they had decided to turn back. After passing them, I saw there were no longer any other climbers in front of us, so now we had to break trail. At this point we were climbing in pretty deep snow, which means the front person has a much harder job than the person following in the tracks. As a result, Rémy and I changed lead position a number of times. I found it not only physically demanding to create a path through the snow but also challenging to find the best route across rock formations. If you go up the wrong way, you have to climb down again and find a better way.

When I looked back I could still see a couple of other climbers behind us in the distance who were going up, but Rémy said their guides would soon turn them around as they would have no chance to make it to the summit and back before nightfall. He added, "No one has been able to climb the Eiger in weeks, and at this stage only you and I have a chance to make it to the summit. Let's do it!"

Looking back at the ridge we just climbed. The climbers on the left would soon turn around.

As we went along the ridge that by now had become very steep and was almost completely covered by snow, I had to rely on my crampons and my ice axe. I wouldn't be able to hold on to anything with my hands if I fell. Rémy also warned me that we were now climbing across a big corniche, an overhanging edge of snow, which meant I should be careful not to break through the snow on this ridge, as it would most likely collapse. This made me very concerned, but we managed to carefully make our way across this corniche, which was also the end of the ridge, only to find out that an even bigger challenge awaited us.

Not only was the scrambling going to be more difficult as the limestone rock had made way for granite rocks, which are much looser and hence don't provide the same hold, but also we had to traverse a near-vertical ice field which stretched at least 50 metres. Rémy went first and safely made it to the other side, while putting in some ice screws for protection during his traverse.

When I started to cross, I immediately felt that my left crampon started to come loose. There was no way I could tighten it while clinging on to this ice wall. I knew I was putting a lot of pressure on it as I was kicking hard into the ice to get a good grip. I could only hope that my crampon wouldn't come off. Without crampons I would not be able to make my way across or turn back. I would likely slip and would have to hope the ice screws that Rémy put in would hold the rope attached to my harness. Not a good prospect. On top of that, my hands started to get very cold as I had to take out the ice screws every time I passed one.

When I was halfway across this ice wall, I could see what I would encounter beyond this face. Fortunately it would only be a couple of pitches on relatively steep rock to the summit. I estimated that it wouldn't take more than an hour, which gave me hope.

When I finally reached the other side of the ice field, with that crampon almost falling off, I started to believe we could probably make it. I tightened it again and got myself mentally ready for the last part of the climb, which is actually on the north face. The climb up from where we were was not too demanding, as I had foreseen, but when I looked down I realized how daunting this wall of death beneath me was. It was a sheer vertical face of almost 2 kilometres! I forced myself to look up and was climbing those last pitches with great excitement as I was now climbing the very north face of the Eiger! I was relieved when I finally made it safely to the top at 11.30 a.m. and could barely comprehend it.

On top of the Eiger, a very proud moment.

From the summit I could see the Mittellegi Hut, from where the alternative route starts. No one was climbing the Eiger from there either, so Rémy and I were very privileged to have had this legendary mountain to ourselves that day. I also wanted to take a picture of the north face now that I was safely standing on the summit, but

Rémy explained to me that I actually wasn't safe at all. "TJ, don't move. You're standing on top of a corniche, and if you take one more step forward, it will break and you'll go all the way down the north face. Slowly make your way back towards me." This was bloody frightening, and I was very happy to leave the summit after a couple of minutes. Rémy didn't need to remind me that we were only halfway, and I was fully concentrated as I remembered my fall during the descent of the Matterhorn only too well.

There are no real possibilities to abseil on the descent, apart from the first bits. After that it really involves climbing down, and up and down again over the ridge. I found this very tiring as I'm tall and hence my balance isn't great. Climbing down took me over seven and a half hours, whereas the ascent had taken me six and a half hours. So after a gruelling fourteen hours, we made it back to the hut just before 7 p.m. I was relieved and thankful to have summited and made it back safely.

Rémy and I celebrating our success back in the hut.

None of the other twelve climbers who had attempted the Eiger that day had gotten close, but they all came over to congratulate us, which I greatly appreciated. I was also very thankful that Rémy and I had been climbing in perfect weather. Not only had the two previous weeks been bad, but also the following week would be horrible. I realized how incredibly lucky I was to have climbed the Eiger on this particular Sunday. This really felt it was meant to be.

Now that I had climbed the big three in the Alps, I started dreaming about other mountains in the world I would like to climb. Definitely try Denali again, and maybe Everest one day? But I didn't want to get ahead of myself and made sure to savour this particular moment. I ordered a bottle of Pinot Noir in the hut to celebrate our success but explained to Rémy that I would still order that bottle of Cheval Blanc as promised once we got off the mountain.

The next day Rémy would be driving me back to the train station because I had to get back to London that evening as I had an important business trip the following day. When we got to the car park in Lauterbrunnen, I insisted we first have lunch at a restaurant where they served fine wine and started to google good restaurants. As the town of Vevey was on the way, I looked at a number of different restaurants and decided to call the Trois Couronnes. The photos on the website showed the beautiful views overlooking Lake Geneva, and the restaurant looked reassuringly expensive. *Surely they must have a good wine list too*, I thought.

When I called them to ask whether Cheval Blanc was on the wine list, I was put through to the sommelier, who was happy to confirm he had that wine from the 1999 vintage. It was still a bit young, he explained, but perfectly drinkable nevertheless. I told him we would be there in an hour, and he asked for my last name, kindly offering to make me a reservation.

When I told Rémy that we were on for that promised bottle, he looked very happy but also very surprised. "Rémy, you look surprised. Did you think I would forget about the Cheval Blanc that I promised you?"

He simply said, "No, I didn't think you'd forget, but you're Dutch, and that means that you're a cheapskate!" You have to admire his honesty.

When we got to the restaurant, it turned out to be part of a five-star hotel. When Rémy drove his old banger onto the parking lot, it clearly stood out. The even bigger issue was that we had been in the mountains for days without washing, and the only clean clothes I had were a pair of jeans. And Rémy didn't even have that! I told him to use lots of deodorant and walk into the place as if he belonged. "You go first," he said.

When we approached the maître d', she was not pleased to see us and looked us up and down. I greeted her with a smile on my face and a very polite "Bonjour, Madame, comment-allez vous?" She clearly assumed we had come to the wrong place.

"Well, eh, gentlemen, can I help you?"

"Very much so," I replied. "We have a reservation. My name is TJ Halbertsma."

All of a sudden she turned appreciative of our presence and replied very politely: "Ahh, Monsieur Halbertsma, mais soyez le bienvenu!"

She asked us to follow her to the terrace. The view of the lake was breathtaking. She offered us the best table, right in the middle. When we sat down, Rémy had to giggle. I have to say I enjoyed it myself as well. The bottle was already standing on the table, and when the sommelier came over to introduce himself, I saw he couldn't understand that these two climbing bums would order this very fine claret. When I explained the story, he had to laugh. He congratulated us and immediately opened the bottle for us. It had to breathe a little before we could drink it. After tasting it, Rémy said it was so good that he wouldn't even have anything to eat as the food could interfere with the taste of the wine. The bottle started to get better the longer it breathed, so we really had to pace ourselves, which wasn't easy after having been on the mountain for days. But it was very much worth it. What an incredible bottle of wine it was. The price was also pretty extraordinary (1,100 Swiss francs!), but then again I never would have stood on the summit of the Eiger without Rémy.

It was two years later, in 2013, when I was climbing Mount Rainier in Seattle, that I shared my Eiger adventure with my guide. He told me I had been fortunate to have climbed with Rémy, and I realized that my friend had a reputation that stretched far beyond the Alps. Although my guide had never met Rémy, he had followed his exploits for years. I said, "I'd like to go climbing with Rémy again next summer," which was when his face turned grim. "Haven't you heard what happened last year?" His expression made me assume this was not good news. "Rémy was caught in bad weather on Manaslu in the Himalayas and perished in an avalanche on 23 September 2012."

This news hit me hard as those days that I had spent with Rémy had been very special for me. I thought back to my time on the Eiger with him and then realized that buying that expensive bottle of Cheval Blanc was probably the best money I'd ever spent.

Chapter 6

THE HIGH LIFE CONTINUES

I have always enjoyed skiing. In my youth my brother and I went skiing on numerous occasions, and it was my introduction to the mountains. We had lots of lessons, and he enjoyed it as much as I did. When I was in my twenties I started to appreciate the long lunches and après-ski too, and most of my skiing holidays started to be more social rather than full-day skiing holidays. To be honest, at some point you get bored with regular on-piste skiing. I will always enjoy my time up in the mountains, but going down the slopes with thousands of others, queuing for lifts, and then doing the same thing all over again, it loses its appeal at some point. So when I turned 40, I started to look for the next skiing challenge.

An obvious next step was off-piste skiing, but I knew the dangers should not be underestimated. Mind you, it was early 2012, and Friso van Oranje, with whom I was well acquainted, had just been caught by an avalanche in Austria, on 17 February to be exact. Notwithstanding the fact that his mother was the Queen of the Netherlands, he was the most down-to-earth and thoughtful person you could ever meet. When I heard the news that he had been trapped by an avalanche, I knew that it was not the result of carelessness. Unfortunately it had been a dire set of circumstances, and it served as a stark reminder that the mountains don't make an exception for anyone. Sadly Friso is no longer with us. I feel very privileged and honoured to have known him.

My wife and I, together with a group of friends, went on a skiing holiday in Courchevel only weeks after Friso's very unfortunate accident. It had been worldwide news, and everyone was still shaken. I had to promise my wife I wouldn't do anything out of the ordinary, as she knows I like to push my boundaries. She specifically told me to resist the temptation of going off piste to explore the 600-kilometre skiing area of the Trois Vallées, the largest ski region in the world, which Courchevel is part of.

During this holiday none of us went off piste and we managed to avoid any accidents on the slopes. On the last day of our holiday, one of the group, who were all very accomplished skiers, came up with the idea to go ski jumping that afternoon.

Everyone was very enthusiastic until we actually found a ski ramp. The first person of our group to ski jump was a failed financier who very much wanted to prove himself. He was a very experienced skier. In fact, he used to be a ski teacher, so he knew he would be able to pull it off. Full of confidence, he went down the ramp, and his jump was flawless, as was his landing. It looked so easy. I was going second, and my gut feeling told me this wasn't a good idea at all. Nevertheless I set off, not really knowing what was going to happen. The next thing I knew, I was flying through the air. Although my jump didn't seem to last very long, I did go far and high. And the pain when I landed was excruciating. I had fallen over, and I had heard a loud crack upon impact. All the spectators were applauding, but it was clear to me that something was very wrong as I was in terrible pain.

I slowly got up. My immediate assumption was that I had broken my collarbone because any shoulder movement was simply unbearable. I wanted to go down as soon as possible, but I knew that skiing to the village was not a good idea. I didn't want to be flown out by a helicopter or be helped down on a rescue toboggan and felt very embarrassed. I told everyone I was OK but said I would call it a day and made sure to ski down very carefully.

I walked from the slopes into town in terrible agony. One of my friends joined me as he knew that things were not all right, which I appreciated. We had to wait for about half an hour before Doctor Pepin would see me. When he examined me, he said, "I don't even have to take an X-ray. It is very clear what the issue is as your collarbone is sticking out and pointing in the wrong direction!" My suspicion was confirmed there and then. "The good news is that I don't have to operate on you as it seems to be a clean break. The bad news is that I have to reset your clavicle for it to heal properly and avoid a malunion."

The doctor didn't tell me specifically why that was bad news, but I quickly found out. Two nurses were holding me, one on each side, as the doctor approached. He put both his hands on my left shoulder, one on the front and one on the back, and I asked him how he planned on aligning the two bones as I had not had any anaesthetic. Next thing I knew, he pushed as hard as he could to straighten my collarbone, which was undoubtedly the most painful experience of my life. I now understood why those two nurses were holding me; I almost fainted. They put me on a chair and wrapped up my shoulder to make sure my clavicle would stay straight. I was still dazed from the pain when I put on my skiing fleece and was given a sling to hold my arm, after which I was free to go.

I had been told by the doctor that the healing process would take only six weeks and that I would be fine again after a couple of months of physiotherapy. But it was evident to me when I left his practice that day that I would never go ski jumping again. When I got back to the chalet we were staying at, you can imagine that my wife was not pleased. I had to look beyond ski jumping if I wanted a new challenge in the mountains. I then remembered that Olly, my regular climbing guide and by now good friend, had once said, "If you like mountaineering, you'll also like ski touring." It involves "uphill skiing" whereby you attach skins to your skis that allow you to ski up a mountain without gliding down. It obviously also involves downhill skiing, for which you simply take off those skins. And you'll also be climbing mountains with your skis on your back. It was in 2014, when I had a reunion with a group of old friends, that I started to consider ski touring seriously.

They were reminiscing about their latest ski mountaineering adventure, the Haute Route, and asked me whether I would ever consider doing it. Mind you, the Haute Route is the most famous ski tour in the world, from Chamonix to Zermatt, the two historic capitals of mountaineering. The scenery is spectacular, starting with views of the Mont Blanc and finishing with a long downhill ski along the Matterhorn. This high Alpine traverse is 120 kilometres long with a 6,000-metre ascent and descent. So obviously it is a very special undertaking, but you have to be a pretty good off-piste skier, which I wasn't. To be honest, that is why I hadn't considered it. Yet. Hearing their stories, I found the Haute Route very appealing: a great combination of skiing, climbing, and scrambling in the most beautiful surroundings.

Not much later I called Olly to reconnect and enquire about the Haute Route. As expected, he very much encouraged me to do an adequate amount of preparation before embarking on a serious tour like that. He explained to me, "You have to be able to ski off piste for a week and make dreaded kick turns on steep and icy slopes in order to change direction. Furthermore, you need to deal with cold weather, strong winds, crevasses, and potential avalanches. But you'll love it."

I told Olly my on-piste skiing was all right but my off-piste skiing abilities were nonexistent. He convinced me to do a four-day ski touring course to get to grips with the basics of off piste, which I did in Chamonix in December 2015. Not only did I improve my skiing on that course, but also I learned about the safety trilogy: the combination of an avalanche probe, a snow shovel, and a transceiver, which are indispensable in case of an avalanche. I complemented that with three more training weekends with friends in the following months, so by April 2016 I was ready and left for Chamonix to start my Haute Route adventure. The Haute Route is normally done late in the skiing season as the crevasses will be filled with snow and most of the avalanches will already have occurred earlier in the season.

Getting the hang of it in Chamonix in 2015.

When I got to Chamonix, Olly informed me that the weather wouldn't allow for the Haute Route. This was a big setback, but I learned this is par for the course. The weather needs to be absolutely perfect to even start the Haute Route: no fresh snowfall, as that will increase the danger of avalanches, and good visibility to make sure you won't ski into a crevasse. But even if you start with great conditions, you have a less than 50 per cent chance to complete the tour because you are very likely to encounter bad weather along the way.

We went ski touring in Italy instead, which was great. I got to perfect my kick turns, further improve my off-piste skiing, and climb a number of peaks, including Gran Paradiso, which I had climbed before in the summer of 2008. This was a different experience though, as the temperature was now −20°C. I certainly wasn't dressed for it. Rarely have I been so cold. But the whole week was amazing and set me up perfectly for the Haute Route at some point in the future.

Olly told me that this ski tour had been technically and physically more demanding than the Haute Route, which I was pleased to hear as it alleviated doubts about my skiing abilities to complete the Haute Route. I became even more committed to do it and decided to give it another go the year after, 2017.

Unfortunately I didn't have a lot of time to train in the mountains, but I kept fit in the gym and managed to squeeze in one day of training in Verbier, although I had hoped to do more. Nevertheless, I felt up to the task when I arrived on Saturday, 8 April 2017, in Chamonix. I met my good friends Arjan and Leo, and we had planned to go for a warm-up ski the next day in the Vallée Blanche, the world's longest off-piste ski route. You take the lift to the Aiguille du Midi at 3,842 metres, which I knew well from my Mont Blanc climb, and then ski down 20 kilometres, which would certainly get me in the right mindset for the Haute Route given the glaciated and crevassed terrain that follows all the way down.

I had skied with both Arjan and Leo before and knew they were good skiers. Arjan is tall, strong, and very determined in everything he does. He had trained hard for our adventure, and I knew he was capable of completing the Haute Route. He's good fun too, so he would be great company. Leo, whom I had also been skydiving with, is always in for an adventure, and you can also have a laugh with him, so we were definitely going to have a good time. Leo had only decided last minute to join and hadn't had the opportunity to get as fit as he would have liked. He's a great skier too, but he still had a bit of extra weight to carry, which wouldn't really be an issue skiing downhill, but it would probably bother him skinning up the mountains. Fortunately the Vallée Blanche is only downhill. We all enjoyed the most spectacular glaciated scenery with views of many peaks of over 4,000 metres.

The Vallée Blanche Route ends in the Mer de Glace, where I had had my crevasse training when we were preparing for the Mont Blanc climb almost twenty years earlier. I could hardly recognize it anymore because so much of the glacier had melted away. When we walked up the path to the train station from the Mer de Glace, I could see the plaques on the rocks indicating the ice level of the glacier year by year. It clearly showed that this glacier is melting away. What worried me the most is that the distance between those plaques was getting bigger every year.

That evening we met Olly and four Scottish friends who had signed up to do the Haute Route with us. This time Olly had good news as the prediction for the next day was clear blue skies without any snowfall. The forecast for the rest of the week was equally perfect, and we were all thrilled to hear it was a go!

With no time to waste, we set off the next day for the Albert Premier Hut, which took us quite long to reach, although we didn't ski or skin too much. We started at the Grands Montets lifts, where we had to wait for ages to get on, which was another reminder that on-piste skiing with the crowds is not my favourite thing anymore. But once we got off piste at the Argentière Glacier, the adventure had truly started as we had to ski, skin, and use our climbing crampons. Everyone was well-accustomed to skiing down and skinning up, but the crampons were a relatively new experience for some of the team members, as we found out on the Col du Passon. Leo in particular hadn't had much experience and didn't find it easy to get a good grip on the snow with those spikes under his ski boots. It was a warm day, so the snow was soft, and often he would sink down to his middle. Fortunately he didn't take it too bad, he had a great attitude and I had to think back to our skydiving days when his instructors needed weight bags to keep up with him after jumping out of the plane.

Leo and I having a laugh. Extra weight can slow you down.

After seven hours we finally got to the hut. I'm pretty much used to life in huts given my mountaineering experience, but for Leo and Arjan, this was a new experience. They're both successful financiers from London and normally stay in good hotels when they go on holiday. Here they had to share a room with many others and be subject to everyone's snoring, including mine. They also realized they'd be lucky to find running water. And the WCs were uncomfortable to say the least. Plus, we were at altitude which gave them headaches, but they handled the discomfort very well. Only six more nights to go! The next day we started early. In fact, we would start early every day as you want to go out when the snow is still solid and the risk of avalanches is low. We got up at 5.30 a.m., had breakfast, and got ready to set off at 7 a.m. to our next destination, which was the Trient Hut. This was going to be a relatively short day, but we would be skinning on some pretty steep slopes, and the skins we had attached to our skis would no longer be enough to get us up those slopes. Olly told us to start using our ski crampons, also known as *couteaux*. They engage in the snow when you step down but slide up when you move your ski forward. I was happy to have them as I certainly wouldn't have managed to get up these slopes without them.

We got to the Trient Hut in under five hours, so we were pleased to avoid the blazing afternoon sun and relax in the hut having a rösti. Again, we didn't get much sleep given the snoring contest that was going on at night.

Stunning views from the Trient Hut.

By the third day we were well into our adventure and set off with a beautiful downhill ski. Then we had to traverse the glacier with a big rock face on our right. We were following the ski tracks from a previous group, but those stopped all of a sudden, and I couldn't really understand why. Olly told us, "This is the Col des Ecandies, and it is the best place to climb over this rock face.

The Col des Ecandies is very daunting.

Take your skis off and put on your crampons." Were we really going up this face? When I was putting on my crampons, I looked up this almost 100-metre vertical wall of snow-covered rock. This would even put an experienced mountaineer to the test. And we would each have to climb this while carrying a big backpack with two skis and poles attached to it! I could only imagine the anxiety from Arjan and Leo, as this would be the first time that they would be climbing a near-vertical wall.

We were ready to go after ten minutes. Before we set off, Olly said, "Make sure you clip onto the rope I will put in place. Take your time, and trust your crampons." Off we went.

Olly climbed up first to set a belay so our rope was secured in case we fell, but it didn't make it less scary. The first bit was already challenging as there was no snow to hold the loose rock we were climbing on. It's very easy to dislodge a rock that can fall on someone else's head, so we all had to be very mindful of where we were placing our feet and hands. After scrambling up 25 metres, we started to encounter snow, which gave our crampons a good grip and made the climbing a bit easier. But judging by their swearing, Arjan and Leo were not enjoying this.

The biggest issue for Arjan was that one of his crampons started coming off, which I hadn't realized. After half an hour we made it to the top of the col and were all exhausted. Arjan was pleased to have made it in one piece, but he told Olly that his crampon had actually come off 30 metres below. He had been so focused on his climbing that he hadn't even tried to recover it. We could see it sticking out of the snow below. Olly kindly offered to climb down to collect it, but he stated clearly, "You'll definitely need your crampons many times this week, so be careful with your kit." We were all thinking, *How many more of these will we have to climb?*

Climbing up the col where Arjan, second from the top, is about to lose his crampon.

From the top of the Col des Ecandies, we had a relatively easy ski down to Champex. Once in the valley we had to take a taxi to connect to the Châble lift in Verbier. Sadly Leo was not able to continue the adventure because of prior engagements and bid farewell to the group at this stage. We only had time for a quick coffee in this buzzy town, after which we went straight back up as we still had to get to the Prafleuri Hut before sunset. When we got out of the lift at the Col des Gentianes, we had to skin over two more cols and then traverse the Rosablanche glacier before we would get to the Prafleuri Hut. Although the skinning was relatively straightforward, it was getting quite late in the day, so we started to worry about the risk of an avalanche as the sun had warmed up the snow.

During our traverse across this glacier, the mountain face on our right looked very unsettling. It was steep and high, and the sun had been shining on it all day. When we had to ski down a precipitous slope, we knew we had to be very alert. We didn't want to be slow, as big blocks of ice could drop down any second from above, but falling on this very steep slope was not an option either, so we didn't want to go too fast either. Olly went first and showed us the safest path to cross. I followed swiftly, but Arjan decided to take another route by skiing a bit further on where the slope would be less steep. This turned out to be a wrong decision.

Arjan had started sliding down on a snowpack that was now gathering momentum. He just managed to ground himself and fortunately wasn't swept down, but it was close. The avalanche he had caused started to accelerate below him and would have swallowed anyone skiing lower down. Luckily the rest of the group had waited behind and weren't affected by this giant moving snowpack, but this was a clear reminder for everyone that accidents can happen very easily. When we got to the hut at 4.30 p.m., we were glad to be safe but were worn out. What a day it had been!

The following day we set off early again, this time for the Dix Hut. We started with a very steep skin up the Col des Roux with our couteaux on, but conditions weren't great. The snow was very slushy, which made the downhill ski that followed very slow and heavy. We then traversed along the Dix Lake, which turned into scrambling over rocks with our skis on our backs as the snow had already disappeared here. After this we had to skin up for a couple of hours to get to the Dix Hut. The sun was shining brightly, we were all sweating immensely, and it was very difficult to stay hydrated. We all had run out of water well before we got to the hut, and my left foot had also started to hurt badly.

One has to be very careful to avoid blisters as it could mean the end of the expedition. We were all using 2nd Skin Plasters to minimize the risk of blisters, but I got worried nonetheless. When I took off my boots, I immediately checked my foot and fortunately didn't have any blisters, but the rubbing of my foot against the inside of my boot had taken its toll. My foot was swollen, which made the pressing even worse. Furthermore Olly had developed a flu, which explained why he had been slow all day.

Once settled in the hut, we immediately rehydrated by drinking copious amounts of water, and the rösti again did wonders. We felt better after our refuelling, but if anything my foot would start to become a bigger problem. And Olly looked worse for wear when we all went to sleep.

The following day we set out to reach the highest point of the Haute Route. Psychologically this was encouraging as it would mean that we would get more downhill skiing rather than gruelling uphill skinning after that. Olly informed us it would be a tough morning with about 1,000 metres uphill before we would get to the summit of the Pigne d'Arolla at 3,790 metres, the highest point. We set off at a rapid pace, and after skinning up steep slopes for three hours, we got to the Passage de la Serpentine. This was the first moment we got a glimpse of the Matterhorn, which brought back great memories for me. It still looked very far away, but at least this meant that Zermatt, our final destination, was now within reach!

From this plateau we were also looking at the Serpentine, which presented one the biggest challenges of the Haute Route. It is a very steep 100-metre slope which is heavily crevassed and exposed. We were roped up when we started this section with our couteaux on, and here I learned that Arjan actually suffers from vertigo. He doesn't get intimidated easily, but halfway up he simply wouldn't move anymore. Kick turns to change direction were very difficult here given the steepness of the slope and the frozen ice we were skinning on. "Come on, Arjan, keep moving. Standing still is not an option," Olly said.

Arjan's reply was resolute: "No way, Olly. This is bloody insane, especially for me, as I've got vertigo. I'm not going up any further!"

All of a sudden I realized what he must have gone through over the last couple of days, in particular when he was climbing the Col des Ecandies where his crampon had come off. It must have been an absolute nightmare for him to go up and down these steep mountains. But he hadn't mentioned it at all as he's a tough guy, mentally and physically. I was happy that Olly was able to get him to continue to the top of this col because we all wanted to get off this slope as soon as possible. Olly had been a schoolteacher in the past, which must have helped in this situation! Not only was I very impressed by how Olly had handled it, but also I was impressed by Arjan, as I now knew what a big commitment the Haute Route must have been for him. He must have dreaded it for months. Talk about getting out of your comfort zone. Well done, Arjan!

After this col, we traversed along the glacier, followed by a short climb on crampons to the summit of Pigne d'Arolla. It was cold and windy, but we had great visibility and stunning views across the Alps. We were towering over any other mountain in the vicinity, and with clear blue skies the Matterhorn was in full view in the distance. It's always such a great feeling of achievement to reach a summit, and this time was no exception. We all felt energized now that most of the uphill skinning was behind us.

On the summit of Pigne d'Arolla, the
highest point of the Haute Route.

The descent through the Arolla valley was stunning with great fresh powder snow, which made it a very special experience in the most beautiful Alpine surroundings with no one else in sight. It gives great pleasure to ski down a pristine mountain where there are no tracks. When you're down at the bottom, you look up, and the only tracks you see are your own.

Just before the Vignettes Hut, we traversed a steep icy slope with huge seracs overhanging. We had to ski this section fast in order to be as little exposed as possible, which was rather unnerving. But all in all this had been another magical day. At the hut Olly seemed to feel a bit better, but my foot did not. It was hurting badly. I had started to take Nurofen to reduce the swelling, but it didn't seem to help a great deal.

On Friday we started with a short downhill ski, followed by an hour-long skin up to the Col de l'Evêque. When we got to the top, we rested for a couple of minutes, which allowed me to take off my rucksack, eat, drink, and unclip my skis to take a bathroom break. When you unclip your skis, you have to put on the snow brakes, which prevent the skis from moving when they're not attached to the boot. As we were on flat terrain now, I didn't bother, although I knew this was a safety measure that one shouldn't take lightly, as I was about to find out.

When I returned from my bathroom break, one of our team members alerted us to a ski that had just started to slide downhill. I was thinking this must have been a very inexperienced skier making a dumb mistake, as losing a ski could mean the end of your expedition. It is very likely that when a ski starts to glide down, it disappears into a crevasse or slides down to the bottom of the valley. I immediately looked at my own skis and was shocked to find that one was missing. It wouldn't be my ski that was now skiing on its own, would it? Unfortunately it was. But as my ski was gathering momentum, there was nothing I could do to catch it. This would surely be the end of the road for me.

I was incredibly fortunate to see a group coming around the corner and up the col where we were standing. I shouted down. They had already seen what was happening. The group's guide was in front and jumped for my ski but missed. Fortunately a woman 20 metres farther down managed to stop my ski with her boot. This was a very brave action from her as she could have injured herself badly. I thanked her a thousand times for catching my ski and carrying it up.

It is still a mystery to me how my ski had started to slide down, as one would assume that a ski would normally not move on flat terrain. Shortly after we got to the saddle at the top, a French group had also arrived, and it is highly likely that one of them kicked against my ski, probably by accident. I could have avoided this all by putting the brakes on, as Olly was keen to remind me in no uncertain terms afterwards.

After this incident, we skinned up for another three hours to get to the Bertol Hut, which is built on a rocky outcrop. The only way to access the hut is to climb up a 50-metre fixed ladder, which Arjan had been anticipating with great apprehension all week. He had checked this out on YouTube and admitted this had made his stomach turn as he really didn't like heights. Olly made sure to clip Arjan onto a rope alongside the ladder, but it was a scary experience nevertheless as it was highly exposed.

When we sat down in the hut, Arjan realized he also had to climb down that ladder the next day, which didn't make him cheery. My foot now hurt really badly. I could barely walk in my boots, let alone ski, and Olly was still coughing. We weren't in a great mood, but we also knew that we were now very close to reaching our goal. Only one more night and one more day to go, which slowly started to lift our spirits.

Olly told us we could be in for a very striking finale. In particular, we could expect spectacular views of the Matterhorn and the Dent d'Hérens. But we would need good visibility for that, which we probably wouldn't be getting. Overnight clouds were expected to come in, and with poor visibility we would have to abandon our plans altogether. The descent to Zermatt is through the Stockji, Tiefmatten, and Zmutt glaciers, with crevasses so big they could easily swallow a bus. You need to see clearly where you're skiing when you navigate this type of crevassed terrain.

When we woke up on Saturday morning, the weather had indeed started to deteriorate. Dark clouds had gathered and a storm was coming in, but we decided to head out as we hoped we would be able to get to the other side of the mountain, where the weather would be better, before getting caught in the storm. Arjan was very relieved when he got down the ladder without incident, after which we skinned for two hours up the slopes of the Tête Blanche. When we reached the top, we were truly overwhelmed by the views of the Matterhorn and the Dent d'Hérens which presented themselves right in front of us. Although we still had good visibility, the clouds were getting thick and dark, so we quickly took pictures while we still had a view.

Olly in between Arjan and yours truly. The clouds are coming in but the Matterhorn and the Dent d'Hérens are still visible in the back.

From here, there would be no more uphill skinning, only downhill skiing, but our joy was short-lived as it turned out to be a complex journey across the glaciers, with crevasses and seracs wherever you looked. It became very clear to us why you only want to do this descent with good visibility.

Going down the beautiful but dangerous glaciers.

My foot was now hurting so much that my skiing had become almost impossible. I was simply trying to get down the mountain without falling. We managed to make it all the way down to the village, walking the last hour with our skis on our back. We looked slightly out of place when we walked into the glitzy town of Zermatt, among women wearing fur coats, going in and out of expensive shops. We had not showered for a week, but boy, I felt like a rock star.

This had been a clear victory for Arjan as he wasn't sure he would be able to complete the Haute Route because of his vertigo. He had trained every day for months to be in the best shape possible, but vertigo is not something you can train for or make go away. I was proud to share that moment with him when we walked into Zermatt, also on a total high. For me it was also an achievement as the Haute Route had been on my bucket list for some time now. Many skiers around the world aim to do it, but only a very few ever get to complete it.

Limping into Zermatt but feeling great!

Chapter 7

ON TOP OF THE WORLD

The Dutch have always had a keen interest in exploring. In the seventeenth and eighteenth centuries, the Dutch East India Company achieved many successes, among which were Abel Tasman's discovery in New Zealand (Tasmania) and the foundation of New Amsterdam, which now goes by the name of New York. But no other story made a stronger impression on me in my childhood than that of Willem Barentsz.

He was an Arctic explorer who lived in the second half of the sixteenth century and was looking to establish a shipping route from Europe to Asia along the Arctic coast of Russia, a route that is now known as the North-East Passage. He didn't know whether it actually existed but assumed there must be a sea passage as the sun shines twenty-four hours in summer, which would melt any Arctic ice. I was especially intrigued by the stories of his heroic exploits during his final expedition in 1596, when he and his crew discovered many new archipelagos in Northern Europe and Eastern Europe, including a land mass which is now known as Spitsbergen, which is part of Svalbard, the northernmost archipelago in the world.

Their ship got trapped after two months at sea between the icebergs and ice floes at Nova Zembla, the easternmost point of Europe, and the seventeen-man crew had to abandon ship. They survived on the island by hunting Arctic foxes, fighting off polar bears, and using the lumber of their ship to light fires and build a lodge in order not to freeze to death during the Arctic winter. After being stranded on the ice for almost twelve months, their ship was still stuck, and Willem Barentsz realized that another winter in these inhumane conditions would mean certain death.

The crew, suffering heavily from scurvy, built two small boats and set out to sea. They were rescued seven weeks later by a Dutch merchant vessel at Kola Peninsula in Russia. Sadly five men didn't survive the journey, among whom was Willem Barentsz. Although his expedition didn't lead to the actual discovery of the North-East Passage, his cartographic contribution has been very significant as he had managed to clearly map the extreme north-east of Europe, which was invaluable for future voyages. In his honour the sea at which he died was named after him: the Barents Sea.

I remember hearing about these Arctic adventures at school, and I was hoping to one day step into the footsteps of these brave explorers. I was particularly fascinated by the North Pole. It represents something unattainable, something inhospitable but yet so captivating. I had always dreamed of going to the North Pole, but I never seriously considered it until 2009. The financial crisis had just hit. I had lost a lot of money and was reconsidering my priorities in life. Over the last fourteen years I had worked very hard and made so many sacrifices, only to see the money I had made evaporate. I thought I'd better start following my dreams as experiences are something that no one can take away from you. I had already been on many adventures, but I was looking for something different, the next level if you will.

And it was in that year, 2009, when I read about the centennial anniversary of Robert Peary's North Polar conquest in 1909. He is credited as the first person to stand on the North Pole, although not everyone agrees. It's a fact that he got really close, but it was very difficult to determine one's exact location in 1909.

It's impossible to locate the North Pole by using a compass. The explanation is that a compass points to the top of the Earth's magnetic field, also known as the north magnetic pole. Interestingly this point is not the same as the true North Pole, which is called the north geographic pole: the northernmost point of the axis around which the Earth rotates. The north magnetic pole is about 1,500 kilometres removed from the geographic North Pole and moves about 15 kilometres north-east every year. Fortunately explorers in those days had an alternative navigation tool, a sextant, which they used in combination with the sun and the North Star to determine their location. But obviously a sextant didn't give as accurate a position as a modern-day GPS does.

So did Robert Peary really stand on the exact right spot, the geographic North Pole? We'll never be sure. Regardless, Robert Peary is known as the first man in history to stand on the North Pole, the top of the world. Reading up on the Arctic, I also learned that one is not sure for how much longer one will be able to stand on the North Pole as the Arctic ice cap is melting fast. And that is when I decided that this was going to be my next adventure. I would make a lifelong dream come true.

Easier said than done though. When I started researching the options, it became clear that standing on the North Pole wasn't going to be straightforward at all. Obviously the North Pole is brutally cold, polar bears are a constant threat, and climate change reduces the ice levels on the Arctic, which makes it dangerous to travel over the ice. And last but not least, how do you get to the Arctic in the first place?

The Arctic is a big ice floe that is connected to Russian land and Canadian land, so it would be possible to start from the Russian or

Canadian coastline, which would take two months to cover the 800 kilometres to the North Pole. My boss was not going to agree to a two-month leave of absence, and to be honest, I wasn't too keen on it either. But then I came across another option. The Russians operate a polar base on 89 degrees Northern Latitude which they run for scientific purposes.

Sceptics believe that Russia only operates this base, called Barneo, because of the vast amount of oil beneath the Arctic Sea, saying the Russians are looking to lay claim to the Arctic for that particular reason. Legally, the Arctic is no man's land as it is not really land to begin with; it's an ocean which is covered by sea ice. There is currently an international process underway, run by the United Nations, to determine who owns that seabed, but that didn't stop the Russians in 2007 from planting a flag on the bottom of the sea at 90 degrees, i.e. the North Pole. This was not an easy feat as the ocean is 3 kilometres deep, and they must have had a good reason to do that. In any case it makes some people question the scientific motives for operating Barneo, the camp they set up every year when the harshest temperatures abate and then break down when the ice starts to melt in spring. No doubt a costly affair.

Whatever the reason may be for operating Barneo, I read that it could also be used for expeditions, and the Russians would welcome adventurers who were looking to ski the Last Degree, i.e. 89 to 90 degrees. This would still be a distance of 110 kilometres, but certainly not the 800 kilometres if you left from the coastline. In other words, it is a great opportunity for people who want to stand on the North Pole and endure what real polar exploring is about but who can't take off more than two weeks. Clearly this was the best option for me.

The next step was to find a guide who would be willing to take me on this Last Degree expedition to the North Pole. And I found out there are not many. I got in touch with an organization called Beluga

Adventures who were willing to introduce me to Marc Cornelissen, a very accomplished polar explorer who happened to be Dutch like me. He was one of the very few people who had skied both to the North Pole and the South Pole. I was very excited to meet him over the phone, and he sounded interested in taking this project on, but he also explained that the logistics would be costly. He said, "No tour operator flies to Barneo, so the Russians have a monopoly. They can charge you what they want. I have good contacts there, so we will probably be able to negotiate a fair price, but I'd strongly suggest you find some others to join you on this expedition so you can all share the cost."

Although it was never going to be easy to find some people to join, I felt optimistic as I presumed the proposition sounded appealing: a once-in-a-lifetime opportunity to ski to the North Pole in two weeks with one of the most experienced polar guides in the world. Marc had said that I should ideally have a group of six or seven people. With more we would need an extra guide, which would lead to additional costs. I sent an email to twenty of my friends who I knew were fit enough and up for an adventure. The initial reaction was very positive. Who would not want to stand on the North Pole, right? *Will I actually have too many people who want to join?* I thought.

A couple of weeks later I sent another email with some details on the required training regimen, which diminished the initial enthusiasm, but the cost was the real deal-breaker. No one was willing to pay $25,000 for the privilege of spending two weeks in a tent with temperatures of –30°C, bar one. My good friend Marius was attracted to the idea as he also realized this was a once-in-a-lifetime opportunity because climate change would only make it more challenging to stand on the North Pole in the future. His confirmation was very encouraging as it meant that my idea was not totally stupid and could be appealing to other like-minded individuals.

I went back to those people who were still on the fence and was able to convince another two friends to come along. Edwin and Bart were both ex–Royal Marines, and Bart had already had extensive polar training in Norway during his army days and was looking forward to go back. Then Frank, an ex-commando, had to step up too. He is also close friends with Edwin and Bart and simply couldn't bear the thought of hearing them brag about standing on the North Pole and him not being part of it. It was interesting to see that the rivalry between the Royal Marines and the commandos was still very strong among these middle-aged men. This expedition was coming together nicely now, whereas three weeks prior I wasn't sure I would be able to get a team together. Ultimately Marius asked two of his friends to come along too, Rupert and Farzad, and now we had the full team together. It was autumn 2009, and Marc was very excited that he could start the preparations for the expedition.

First he wanted to meet us all and find out what these wannabe polar explorers were made of. We decided to meet up in the Netherlands as most of the team members lived there. It was great to finally meet Marc in person, as we had been speaking on the phone over the course of months. He was very amicable, energetic, and knowledgeable. He explained why he had asked us to come to the beach in Hook of Holland: "I need to make sure you guys are up for this sort of an adventure. A Last Degree expedition is very demanding, and if one of you gets in trouble on the ice, he won't be able to take the bus home. I've brought some sleds that I'd like you to pull through the sand." This wasn't only a good opportunity for him to evaluate us, but also it enabled us to get to know each other as not all of us had met before. We all seemed capable of pulling these sleds, even when Marc decided to up the challenge and fill them with sand. After an hour and a half Marc seemed happy with our level of fitness and stamina and told us we were on! This was a great moment for me as we were definitely going to the North Pole. The preparations could now begin in earnest.

We had another six months to go as the expedition would be in April, after the ice-cold Arctic winter passed. Marc told us to start pulling tyres as part of our training regimen. Pulling our sleds, or pulkas, on the ice would be pretty similar. It's important to build up a strong level of fitness but also to train your back muscles, core, and hamstrings. I managed to find a big tyre and a harness and decided I would go to Richmond Park in London for my first round of tyre pulling. I would also use ski poles, which would help me train my biceps and triceps. Slightly embarrassed about pulling a tyre in a public park, I went as early as possible, hoping that I would be the only one there. I couldn't have been more wrong! It was a beautiful Sunday morning in autumn, and the park was bustling as of 9 a.m., when it opened. And it wasn't only the people who were staring at me; even the deer looked at me as if I had gone insane.

Surprisingly I really enjoyed my rather unusual training regimen, apart from the stares. I was outside in a beautiful park and had the opportunity to get my mind off everything for a couple of hours. Marius and Rupert also lived in London, and we trained together when our busy schedules allowed it. The weather was getting colder too, which would help us prepare, to some extent, for the extreme cold we would be facing on the North Pole. Little did we know how cold −30°C would really feel. We were a bit apprehensive to find out a couple of months later. We were also instructed to start getting our gear together, which was going to be an adventure in itself. Very few of the clothes we would require would be for sale in a high street shop.

First of all we would each need a water- and windproof outer-layer jacket with hood, half long to shield your stomach and back from the wind. We would need to sew a fox fur ruff to the hood for extra face protection. In addition we'd need a balaclava and ski goggles. Furthermore, we required a high-insulation down jacket or parka, thin gloves, and over those the biggest and warmest mittens, similar to what one would wear on Everest summit day. Water- and windproof expedition trousers were also required to doubly ensure our stomachs and backs wouldn't be exposed to the elements. On top of that we required two fleeces as underlayers and a base layer for our legs and upper body. This base layer had to be from merino wool, and our underwear had to be polypropylene to make sure it wicked the moisture away from our skin. That way the sweat would not freeze on our bodies when we cooled down. Similarly we needed vapour-barrier socks underneath our thicker insulating socks.

We would also need to bring a sleeping bag rated -40°C, plus a roll-up inflatable mat, plus a thermal foam mat for underneath. The Sorel boots would be provided, together with the skis and poles, as would the tents, stoves, pulkas, and dried food. Lastly we should not forget an eye mask, as it would be light for twenty-four hours and we would need to get some sleep after all.

Our North Pole adventure started on 30 March with a flight from London to Amsterdam, where Marius, Rupert, and I met the rest of the team. From there we connected to Oslo, Norway, and then the next day we flew to Longyearbyen in Spitsbergen at 1,300 kilometres from the North Pole. This was our gateway to the Arctic. I was already excited to make it this far north as Longyearbyen is located on Spitsbergen, the island I remembered well from the adventures of Willem Barentsz.

Our two-day stay in Longyearbyen allowed us to check out our gear, which had been flown in separately given the sheer volume. We practised setting up a tent and getting the cooking stove to work, which was pretty straightforward, but Marc warned us that in a blizzard on the Arctic the simplest tasks would become incredibly challenging. He urged us to practise the routines with our mittens on, which made everything a lot harder, and this was not even in Arctic temperatures.

We flew to Barneo on 2 April, which gave us some of the most spectacular views I have ever seen. The labyrinth of sea ice north of Norway was just breathtaking with clear blue water surrounding these mountain ridges that were covered in snow and ice with clouds rising above. It also took our minds off what was to come. We knew that the cold would hit us the moment we got out of the plane, so half an hour before landing we started to put our warm clothing on. The landing went smoothly, especially if you take into account that we were landing in a big, heavy Antonov 74 cargo plane on a smoothed strip of snow-covered sea ice. This can only be accomplished by the most skilled pilots, and the weather needs to be absolutely perfect, which it fortunately was with clear visibility.

Beautiful views on our way to Barneo.

When I got out of the plane, the cold hit me instantly. Never before had I experienced −25°C; this was quite something. Although I was wearing my down parka, I wasn't properly protected. After two minutes on the ice, Marc came up to me and told me I had already developed a frostnip, a precursor to frostbite, on my nose. "TJ, I know you're excited to be here, but you have to be careful. Your nose is about to freeze. Cover your face ASAP." This scared the hell out of me as I would be here for at least a week! I have a rather big nose, and the only way to properly cover it was to wear a balaclava. I immediately put mine on. Breathing was uncomfortable, but my face started to warm up a bit. I decided to wear it for the rest of the expedition. My nickname became "the Mask".

The eagle has landed. No mean feat on a strip of ice.

Barneo looked bigger than I had expected. There were bulldozers to clear the snow from the runway, many tents, ski scooters, and plenty of sled dogs. Once in the mess tent, we met a number of other explorers who were also looking to ski the Last Degree. Marc introduced us to a fellow guide, Doug Stoup, with whom he would decide what the best drop-off point would be the next day. A Mi-8

A North Pole expedition requires a lot of equipment.

helicopter would drop us all off at a strategic location on the 89th degree, from where we should be able to reach the North Pole hopefully within a week. The most important factor to take into account would be ice-drift because we were, after all, on a giant ice floe subject to the currents underneath. We had an early dinner and went to the tents that were set up for us already and had a pretty comfortable night's sleep, considering that we were sleeping on a layer of ice with a 3-kilometre-deep sea beneath us.

The next day we took off in the helicopter to be dropped off at the strategic starting point on the 89th degree. The pilot told us he would not be able to land on the ice as he wasn't sure it would be thick enough. This was going be a quick exit. Marc shouted, "As soon as we hover above the ice, jump out, and I will hand you all your gear. Put everything in your pulka as fast as you can, and make sure you get away at least twenty-five metres from the helicopter. And be careful of the rotor blades. Stay low, and then lie on your gear. Make sure everything is covered; otherwise your kit will fly away once the helicopter takes off again. Now go!"

The Mi-8 taking off after dropping us at the 89th degree.

After the helicopter disappeared from sight, it was dead silent. The adventure had truly begun. This was very exciting, but we were also anxious as we knew of the dangers that lay ahead of us. Marc said, "OK, team, our adventure has now really begun. Just to remind you, the ice is a constant danger, as you'll only find out too late if the ice is too thin to stand on. Secondly, the cold is something you shouldn't underestimate. Frostbite can happen very quickly, and if one of you suffers frostbite, he will be evacuated immediately to limit the risk of permanent damage or, even worse, amputation." I was wearing my balaclava already and didn't want to take any risks, but I was the only one. I was hoping that no one would get in trouble. Marc continued: "Thirdly, there are polar bears in the Arctic. Normally polar bears don't go further north than 82 or 83 degrees. They hunt seals that come up for air, and the further north one goes, the thicker the ice is. But because of climate change, you now have open water almost everywhere in the Arctic, and tracks of polar bears have even been spotted on the North Pole." This was a real worry indeed as we all knew that there is no negotiating with polar bears. They'll go for you, plain and simple. Fortunately the two Marines we had in our team were good shots, and both were given a rifle to carry, just in case. If we did meet a polar bear, it would certainly try to kill us, so we would have to try to kill it first.

The first day was relatively short as we had arrived in the afternoon at our starting position and were still getting used to our daily routine of pulling pulkas, setting up tents, cooking, and learning how to navigate. We did, however, encounter our first pressure ridges. These are formed by cracked-open ice that is compressed again, and they look like giant ice cubes. If you're lucky you can ski over the small ones, but the big ones will slow you down tremendously as you have to take off your skis, climb over them, pull your pulka across, clip your skis back on, and then repeat the process at the next pressure ridge. Once you've crossed ten of them in an hour, you start to get really sick of them. And one of the issues was our bindings as they were not easy to operate; you had to take off your gloves to put your skis back on. This severely increased the risk of frostbite.

In the evening we started to realize how long it would take every day to set up camp and cook our food. The most time-consuming activity was the boiling of the snow. The sea ice contains salt, which makes it unsuitable for drinking, so you have to use snow, as dense as possible. The more compact it is, the less snow you need to melt into water. It took us two hours to get one pan of boiling water alone! We also found that one of the stoves wasn't working, which our handyman Rupert fortunately knew how to fix. All in all it took us a couple of hours before the tents were set up and we had dinner. This was going to be our routine for the week.

Many Worlds to Conquer

Navigating our way through the pressure ridges. Not easy!

Crossing giant ice cubes is even trickier
and slowed us down tremendously.

The next day, 4 April, we wanted to set off early, but the temperature had dropped to -30°C, yet with the wind chill factor added, it felt like -40°C. We managed to ski a full day though, and after crossing many more pressure ridges, we were glad we could set up camp and crawl into our tents. The issue was that you stop moving when you set up your tent and then you can get cold very quickly. That is what happened to most of us. I felt my right foot going numb by standing still. Although I was freezing, I didn't want to go inside my tent when the others were still setting up their tents, but then my lips turned blue and I started to shiver. Marc noticed this and immediately told me to go inside my tent and get into my sleeping bag. The others were probably very cold too, but no one wanted to admit it. We were real explorers, right?

Farzad and I battling the brutally cold Arctic weather.
I'm happy to be wearing my balaclava at -40°C.

The following day it became apparent that we had an issue. Edwin had suffered frostbite as he had taken his mittens off too often the previous day, and it didn't look good as three fingertips had frozen.

Marc assessed the situation and quickly realized we had only one option, and that was to organize an evacuation for Edwin. Marc used his satellite phone to call Viktor Boyarsky, the Barneo manager, to request help. Viktor's reaction was rather muted. "Tell me, Marc, how bad really is the frostbite? Is it the whole hand or just a couple of fingers?"

Marc replied, "Well, it's definitely a couple of fingers, and you can clearly see damage on the fingertips."

"Oh, just the fingertips? Are they black, or are they blue?"

Marc had not expected these questions but had no other option than to politely answer them. "They have definitely turned blue but aren't black."

"Marc, let me get this straight, three fingers only and not even black? I'll see you next week. Bye-bye!"

And that was it, no pickup for Edwin. The Russians probably didn't want to "waste" fuel unless it was a life-threatening situation. It became clear that their definition of emergency was very different from ours.

After that we installed a new rule, namely that Edwin was not allowed to take off his mittens for the rest of the expedition. And we would do everything for him: close his bindings, zip up his parka, set up the tent, etc. The only thing we didn't help him with was the call of nature.

One might wonder how you go to the bathroom on the North Pole. The answer is, quickly! You simply have to go behind a pressure ridge and do your business, ideally out of the wind as you'll literally freeze your butt off. I presume Edwin wouldn't have appreciated any help anyway. We also had bottles we would use if we had to pee at night, as it would be very uncomfortable to get out of your sleeping bag and put on all your thick clothes to go outside. Marc had offered sound advice here: "Make sure you don't confuse your pee bottle with your water bottle."

The fact that we were taking care of Edwin had a very positive effect on the group. We were already cohesive as we were friends before, but this was a real bonding experience which would help us get through what was yet to come.

On 6 April the weather was bad, in particular the visibility. We had a complete white-out, which made navigation nearly impossible. We could no longer distinguish the terrain from the sky, and distances were impossible to read. It could also be dangerous as we constantly had to be on the lookout for open water which is ice that has cracked open. We would normally walk around an open lead, but they can be kilometres long, so if we were lucky, open water would refreeze to the extent that we could then ski over it again. And that is normally the best ice, as no pressure ridges have yet been formed.

At one point we reached open water that had just refrozen one or two days earlier, and the ice was so smooth that it was almost a highway. We were very lucky that the direction of this refrozen stretch was north, but obviously we would have to be very careful as the ice was still thin. We were able to ski on this flat stretch for about an hour, during which we covered good distance. We wanted to see how much progress we had made and checked our GPS, only to find out that effectively we had been standing still.

How was this possible? It turned out that the ice had started to move against us and was pulled in a southerly direction by the currents underneath. This was a real downer, but there was not much we could do about it apart from skiing on. When we checked our position again

Many Worlds to Conquer

Encountering open water. How can we cross this?

at the end of the day, we had barely made any progress all day, which was a great disappointment as we had worked incredibly hard in the toughest conditions for almost twelve hours.

When we woke up the next day, we were hoping that the negative drift had stopped, but when we rechecked our position, we saw that we had lost another 3 kilometres. The day before we had worked so hard, but our progress had not been commensurate. In fact, we were now farther away from the pole than we'd been twenty-four hours ago. All that hard work had been for nothing.

We didn't have time to waste and held a crisis meeting. We called Barneo and learned that the negative drift was going to continue. They said we should expect that overnight we would lose at least half the distance we would ski during the day. This was not very motivational, and it didn't seem likely we would reach the North

Would the ice be strong enough to hold us?

The refrozen ice, strong enough and without pressure ridges, allowed us to make good progress. Or so we thought.

Pole on skis by our scheduled date. The pickup was in three days, and we had neither the time nor the food to extend our expedition by a week or more. And an extra week is what we would have needed, if not more.

We were now faced with two choices. Either we would give up and ask for a pickup, or we would continue to ski as far as we could and hope for a push on the last day from the Russian helicopter to the pole. Frank, the ex-commando, had lived up to our expectation of being tough. The cold didn't seem to bother him that much. He didn't get tired and was helping Marc to navigate when he needed a break. He was also vocal during this discussion and had established a lot of credibility over the last couple of days. He said, "If there's one thing I've learned at the commandos, it's that your body can do much more than you think. I guarantee you that we will be able to ski nonstop for forty-eight hours or maybe even seventy-two hours, which will give us a chance to make it after all."

Rupert seemed less convinced, saying, "Even if we did ski nonstop for days, we still wouldn't make it, as the negative drift hasn't subsided. It like's walking on a treadmill—bloody pointless."

Marc did a great job of engaging everyone in the decision-making process and then took the final decision. "We all want to get to the North Pole, but safety is my absolute priority. If we ski nonstop for days, we'll be too tired to focus, and that's when accidents happen. We'll ski on, but no more than twelve hours a day, and we'll see how far we get. When the Russians pick us up, they might even fly us to the pole, but nothing is guaranteed." This was the moment when we all had to accept that we were unlikely to reach the North Pole as planned. We all went to sleep disappointed.

Our expedition had been given quite a bit of media coverage in the Netherlands, and the next morning Marc was supposed to give an update on our adventure by satellite phone for a Dutch breakfast TV program. The presenter was Arie Boomsma, who was very popular, so I knew that many people would be watching his show. Marc was calling in from the tent that he and I were sharing, and it hurt when he clearly stated on Dutch national television that we were not going to reach the North Pole. I felt low, and when we started skiing I noticed the others weren't in a good mood either. The one thing that we were happy about was the weather conditions with temperatures of only −15°C.

Not long after setting off, we started to encounter more and more open water. We managed to circumvent it in most instances, but we also thought it would save time to cross the narrower open leads directly by using Marc's pulka as a bridge. We decided to give it a try when we couldn't manage to find a way around an open lead that didn't look too wide. Marc bridged the open water with his sled, then went across it. Indeed it looked straightforward: climb onto the pulka, slowly move forward, and get off again on the other side. No one in the group volunteered to go second, but then Marius, not afraid of a challenge, said he wouldn't mind going next.

He started off well but was too quick to jump off the sled onto the other side. He didn't quite make it and dropped into the water down to his waist. He was still able to hold on to the sled with one arm and on to the ice with his other arm, but this was a very serious situation indeed. He could be pulled under the ice by the current, which would mean certain death. Furthermore, the ice-cold water would numb him within minutes, which would make it impossible for him to climb onto the ice again. We had to act quickly.

Fortunately Marc, who was already on the other side, was able to grab hold of the arm with which Marius was holding on to the ice and tried to pull him up. At the same time we were trying to drag him onto the ice on our side and started hauling the sled towards us. Marius was now being pulled in two directions and quickly took charge. He shouted at us, "Stop dragging the pulka away. I need it to push off from!" After half a minute he managed to get onto the ice with Marc's help, but this had not been easy as the ice was very slippery. It was very fortunate that Marc had been able to help him.

When Marius got out of the water, the problems still weren't over. As his legs were completely wet, Marius had to change clothes quickly, because otherwise he would freeze up. Marc told us to throw some dry pants and socks from our side and told Marius to start rolling in the snow to keep his circulation going. Marius kept his cool. When we tossed Marc some spare clothes, Marius did a "striptease" on the ice and changed as quickly as he could.

Fortunately it was a mild day, which meant we didn't have to set up a tent, put Marius in a sleeping bag, and prepare him hot drinks. We just carried on, and he was all right. Later that day, when Marc talked to the Russians for our daily update and position, he learned that a couple of days earlier a Belgian team also had someone fall into the water. Unfortunately that had been on a very cold day, and the woman would very likely lose a number of toes to frostbite. Marius had been lucky. What a day it had been. We decided to go to bed early and sleep in.

When we woke up, we went through our daily morning routine, after which we decided to work on a project that Marc was very keen on. One of the things that he had asked us at the start of the expedition was to contribute to a climate change project he had initiated. Mind you, there is no other area in the world where global warming has a more dramatic impact than the Arctic. Over the last forty years, the average thickness of the Arctic ice has declined from 4 metres to 2 metres, and it is likely that within ten years the North Pole will have no ice in summer. The explanation is that an increase in temperature leads to more sea ice being lost, which in turn leads the open ocean to absorb more heat, which in turn will lead to more sea ice being lost. So it's a vicious circle.

What Marc had asked us to do was to help with the collection of scientific data, including measuring the thickness of the ice. This was going to be the day where we could gather good data sets that would be handed over to Christian Haas of the University of Alberta and the European Space Agency for the validation of climate change data that was collected by the recently launched satellite CryoSat-2.

In addition I had raised £23,000 for a weather station in aid of the World Wildlife Fund Arctic Program that Marc was also involved in. That weather station would measure temperature, air pressure, sea ice elevation, and ice-drift (a phenomenon all too familiar to us now) on the North Pole. These scientific activities gave our expedition a special angle as we had witnessed the effects of climate change, in particular the frequency of open water, first-hand.

On 10 April we heard we were going to be picked up by the Russians, most likely around midday, so we gave our position and had to stay put. When the helicopter landed we knew we were one step closer to civilization, but it was also the end of our adventures together during which we had endured so much and built up such a strong bond.

When the chopper landed, Viktor jumped out and greeted us. No words were spoken about Edwin's emergency evacuation that Marc had requested, but Viktor probably did feel guilty as he very kindly offered to fly us to the pole before he would take us back to Barneo. Obviously this was an offer we would not refuse! We had skied 140 kilometres, far more than the geographical distance of the last degree of 110 kilometres. But the negative drift had caused us still to be 50 kilometres from the North Pole! So close and yet so far. It was at 4 p.m. on Saturday, 10 April, that we made it to the Earth's northernmost point after all.

Farzad and Rupert got creative to drill through the multiyear ice. I'm ready to take the measurement.

With a little help from our Russian friends,
we finally stood on the North Pole!

During our flight there, we saw that there was open water everywhere on the remaining stretch to the pole. Even if we had taken an extra week for our expedition, we still would have encountered difficulties trying to make it. When we got out of the helicopter I was proud to stand on top of the world, where I surveyed the place of my dreams, hostile yet so majestic.

A year earlier I had committed myself to standing on the North Pole, and this was the moment that that dream became a reality. Of course I was disappointed that we hadn't managed to travel the full distance on skis, but in the bigger scheme of things, I was very happy and proud to finally stand on the North Pole, something that fewer than fifty people a year achieve.

When we arrived back at Barneo, we heard that Sir David Attenborough was there as well as he was filming his series *Frozen Planet*. He had been stuck at Barneo for a couple of days as a crack had

On Top of the World!

appeared on the runway, and as a result no aircraft had been able to land or take off. Shortly after our arrival we were lucky to hear that a plane would soon be taking off back to Spitsbergen, obviously also to the delight of Sir David. We were fortunate to have left Barneo when we had because the extent of the danger of staying on the ice became clear a couple of hours after we had taken off: a 20-metre gap had appeared in Barneo where we had been sleeping.

During our flight back to Spitsbergen I actually had the pleasure of sitting opposite this legendary biologist and filmmaker. It was a real privilege to discuss a range of topics with this great man, including climate change. Interestingly he was very rational about his North Pole experience and the threat of polar bears becoming extinct. He clearly acknowledged that climate change will alter our ecosystem in many ways. He cares greatly about our planet and is of the opinion that global leaders are not doing enough to address the issue. But he didn't get emotional about that. He approached the topic of climate change very factually and with a great deal of data.

On the plane back with Sir David
Attenborough, what an honour!

Although many people only know David Attenborough as a TV naturalist, here I clearly saw his scientific side. I was blown away by his knowledge on our environment, and it became obvious he is more than a world-famous, eloquent and charming broadcaster. He is also an absolute authority on environmental change. I hope his counsel will be sought by those who can influence climate change at the highest level.

When we got back to civilization, we all had trouble adjusting to normal life. It might sound odd, but we had been so removed from everyday life that worries like paying the mortgage or the Wi-Fi not working seemed irrelevant, as if we had just been on a different planet. I talked to Frank a week after we'd come home. Previously I thought that nothing would ever faze him, but even he said he had difficulty getting back to the normal routine. And importantly,

Edwin started to get some circulation in his fingers again despite some dead tissue. Fortunately, no amputations would be required.

After an adventure like the North Pole, you start to realize what is important to you, like your family and your friends, and that little things can give you ultimate pleasure, like a cup of hot chocolate once you're inside your tent. What a life-changing experience this had been.

Epilogue

Following our expedition, we had yearly reunions as a result of the strong bond we had developed on the pole. Marc would always attend and was never short of ideas for new expeditions, like the South Pole. He continued his polar exploits, but unfortunately I couldn't join him on any given my work schedule. I always felt honoured he invited me though.

He also continued his data gathering for the University of Alberta and the European Space Agency. Just before he finished the project, five years after our adventure, disaster struck during an expedition in the Canadian Arctic. On 29 April 2015 the Canadian Coast Guard received a distress signal from Marc and fellow explorer Philip de Roo. An immediate pickup was requested, but given the bad weather the Canadians couldn't get to their location. While researching the impact of climate change on the levels of sea ice, Marc and Philip had fallen victim to its effects when the ice underneath them opened up. Sadly neither Marc nor Philip survived.

I had to think back to the moment when Marius fell into the water and the difficulty he had getting back onto the ice. It became even more apparent how dangerous that situation had actually been. I never met Philip, but it was a great shock to learn that both men had drowned. I feel very privileged to have known Marc and to have shared many remarkable moments with him. He'll remain an inspiration to me forever.

I felt very purposeful in setting up a trust for Marc's daugther, Jill, to make sure she could continue her education. Everyone from our expedition contributed, and she has shown the stamina, resilience, and determination of her father throughout a very difficult time. This year she's starting her college degree. I know she would have made her father incredibly proud. You go, Jill!

With Marc in our tent on the North Pole. I feel very privileged to have known him as a friend.

Chapter 8

THIS ONE'S FOR MARC

When it comes to legendary exploring, the conquest of the South Pole immediately springs to mind. At a time when almost every region in the world was already discovered, including the North Pole, the South Pole became the last frontier for many explorers, especially for the Englishman Robert Scott and the Norwegian explorer Roald Amundsen. They both had attempted to reach the South Pole a couple of times, and their rivalry came to a head in 1911. There was more at stake than personal pride; this was also about national pride.

The task that lay ahead of them was hard to imagine: before they could even start their journey on Antarctica, they would have to sail the roughest seas for eight months to get to Antarctica, after which they would have to travel 1,400 kilometres over ice in the coldest conditions on the planet. Once they reached the South Pole, they would be halfway only as they would still need to get back.

On Antarctica, Scott would rely mostly on ponies for the five-month South Pole journey he was about to embark on. Amundsen would use dogs for his expedition, which would turn out to be a major advantage in that dogs are much better at coping with the cold than ponies are. Furthermore the Norwegian would be taking a new, more direct route.

When Scott reached the South Pole on 17 January 1912, he was bitterly disappointed to find out that Amundsen had beaten him by thirty-four days. The Norwegian had left a tent on the South Pole and a letter addressed to King Haakon, which Amundsen requested that Scott deliver in case he didn't make it back.

To make matters worse, Scott, utterly exhausted, was running low on provisions. On the return journey he and his men began to starve and had to face the harshest blizzards. They knew they would probably not make it back alive, and one of the men, suffering from terrible frostbite, could hardly go on. He walked out of the tent simply saying, "I am just going outside and may be some time …" He never returned. By sacrificing himself he had hoped to improve the chances of his team members to make it back to safety, but it was to no avail. The rest of the team, including Scott, perished weeks later in their tent at a distance of only 20 kilometres from the depot where they had stored provisions and fuel. Had they reached it, their lives would have been saved.

The story of Ernest Shackleton is another legendary polar story and is often referred to as one of the greatest adventure tales of all time. Shackleton had been on an Antarctic expedition under the leadership of Robert Scott in 1901 and returned in 1908 as the leader of his own expedition. He made it closer to the South Pole than anyone before and gained very valuable experience. After Amundsen and Scott had reached the South Pole, Shackleton's objective was no longer to reach the South Pole. In 1914 he and his crew set off on their ship *Endurance* to cross the entire Antarctic continent with a stop halfway at the South Pole.

Disaster struck when the *Endurance* became trapped in pack ice and was slowly crushed. After ten months it sank, but Shackleton and his twenty-eight-strong crew had already abandoned ship to camp on the floating sea ice until it disintegrated now that winter was passing and spring was arriving. While the ice started to thaw, they had the opportunity to launch the lifeboats. They eventually reached Elephant Island in April 1916, having already been on the ice for fifteen months. With no chance of contacting the outside world for help, Shackleton knew he had to take a risk to save his crew and himself.

He set off with five men in a small boat, the *James Caird*, to look for help. They were aware they would face gale-force winds and 20-metre waves. With only a sextant and a chronometer, especially in these ferocious conditions, it would be impossible to navigate. But it was their only chance of survival. Staying at Elephant Island would mean certain death.

During what would become a truly heroic voyage, ice began to form on the *James Caird*, which started to make it sink. The crew had to work incessantly to chip away the ice, braving hurricanes and ice floes, but miraculously the little boat stayed afloat for sixteen days and 1,300 kilometres, after which they finally saw land when South Georgia appeared before their eyes. There was still no sign of help when they reached the island. They now had to trek across an uncharted mountain range, filled with glaciers and crevasses, to the other side of the island. They did ultimately reach a whaling station and managed to organize a rescue operation for the remaining men who were still stuck on Elephant Island. Not one single expedition member died despite their having been stranded in the harshest conditions for over three hundred days.

After our expedition to the North Pole in 2010, we had also started to think about the South Pole. We had all read these accounts of Scott, Amundsen, and Shackleton, and we were keen to follow in their footsteps. Marc would be happy to guide us again, and he had already started looking at the logistics. However, we were now all back at work and reunited with our families, and to take another couple of weeks off for another polar adventure would be a big ask. We pushed out the idea until the right moment would come along. But of course it never came as there will always be an excuse not to go.

This all changed when we met again at the memorial service for Marc in 2015. We talked about what Marc had meant to us, saying that he had been such an inspiration for us all and that he always told us to follow our dreams. There and then we decided to go for it. Life is too short! We would go to the South Pole in honour of Marc, as he would have been our guide. I immediately started making enquiries.

I quickly learned that the logistics for the South Pole are much more complicated than for the North Pole as the former is much farther away. This would also translate to much higher costs. Furthermore the average temperature during the year on Antarctica is -49.5°C, significantly colder than the North Pole. And last but not least, the altitude would make things even tougher, with the South Pole at an altitude of 2,835 metres. Although that might not seem very high, the effect of the altitude is intensified by the fact that the South Pole is as far removed from the equator as possible, which reduces the air density further. So this would be an even more challenging expedition than the North Pole.

On a more positive note, I also learned that we wouldn't have to prepare for polar bears. The only wildlife we could encounter would be penguins. In addition we wouldn't have to be afraid we would fall through the ice, as Antarctica is a solid land mass. And no, we wouldn't be standing upside down on the South Pole!

Despite everyone's initial enthusiasm, it turned out to be very difficult to get the team together again. What probably hadn't helped was the fact that we hadn't succeeded in reaching the North Pole on skis in 2010. My feeling was that no one wanted to run the risk of a similar experience, especially if the costs would be much higher.

This was very disappointing, but I could understand where everyone was coming from.

So I was back at square one. It almost felt like the North Pole all over again. I had a desire, a determination to stand on the South Pole, but I had no team and no guide. I then remembered that a friend of mine, Serge, had asked me in 2011 about my North Pole expedition. He had seen a presentation of mine, after which he said he was also keen to go to the North Pole. He told me he had already assembled his own team and asked me to introduce him to Marc Cornelissen. Although many people often enquire, this was different; Serge sounded very committed. Obviously I was happy to introduce him to Marc, and when he asked me to borrow my car tyre that I had used for my training, I knew he was serious indeed!

Serge and his team set off for their North Pole expedition with Marc in 2012. Their journey was the polar opposite of what we had experienced two years earlier. They faced mild weather, thick ice, and no open leads—in other words, perfect conditions. Furthermore they encountered positive drift, rather than negative drift, and reached the North Pole in less than a week. Hopefully they would be keen to explore the South Pole too, as their North Pole experience had been so completely different from ours.

When I asked Serge whether he and his team would be interested in another polar adventure, he said he would discuss it and get back to me. I'm pretty sure their successful North Pole journey played an important role in their decision. Serge called me back just a couple of days later to tell me that he and his two friends were definitely game. We were on!

The training schedule for this adventure was pretty much the same as for the North Pole: cardio and strength training, especially for your core and your back, but most importantly pulling tyres in Richmond

Park again! Our clothing requirements were also similar, apart from the fact that we heard we now also needed ski pole pogies, which are insulated pole handle covers like you find on a motorbike. This would ensure an extra layer of protection against the wind and the cold. And everyone would also need to wear a balaclava at all times. Goggles needed to have a venting system as lens fogging had indeed been a problem for me on the North Pole. And on that pair of goggles we also needed a piece of fleece sown on, as a second layer of face protection. Boy, we were in for some cold weather by the sounds of it!

Up until one month before our departure I still hadn't met the other two gentlemen whom I would be going on the expedition with. I was happy that Serge had organized a weekend for the four of us to check our gear, buy what was still missing, and most importantly get to know each other. As I had hoped, Willem and Remko turned out to be great guys: committed, responsible, and excellent fun. And we all had stood on the North Pole with Marc, which immediately created a bond. I couldn't have asked for a better team to go to the South Pole with. We were all incredibly excited about our upcoming expedition.

On Saturday, 3 December 2016, we started our journey, just when the Antarctic summer would begin. Firstly we travelled from London to Madrid, then Santiago de Chile, and from there to Punta Arenas, where we would need to wait for the right weather window to fly the last leg to Antarctica. On Sunday, 4 December, we arrived in Punta and were told we might be leaving already on Tuesday for Antarctica. It would still give us enough time to visit the Nao Victoria Museum and check out the replica of the *James Caird*, the boat that Shackleton used to make that legendary crossing from Elephant Island to South Georgia in 1916, exactly 100 years earlier. When I saw this real-size replica, I could hardly believe a boat this size had been able to withstand those hurricanes and storms that Shackleton had encountered. It made me realize that their survival had indeed been incredible.

With Remko, Willem and Serge, just before
our departure from Punta Arenas.

We also needed to be briefed at the offices of Antarctic Logistic Expeditions (ALE), who were taking care of the logistics and would provide guidance on what Antarctica would have in store for us. This happened on Monday, and sadly we were informed that due to the weather conditions, we would not be leaving on Tuesday. The forecast for Punta Arenas was perfectly fine, but it would not be possible for us to land at Antarctica. The weather forecast at Antarctica needs to be perfect too as turning back is not an option. The Ilyushin 76 that would take us would not be able to carry enough fuel to make it back without refuelling, so ALE was not going to take any risks. Eventually we got the call on Wednesday morning at 7.30 to be ready at 8.00. Our trip to Union Glacier, our base camp at Antarctica, was about five hours, so by Wednesday afternoon we were actually standing on Antarctica. Despite the minor delay, it had taken us only five days from London to Antarctica, whereas the legendary explorers one hundred years earlier spent more than six months at sea before they reached the Earth's southernmost continent.

When we got out of the plane onto the landing strip on ice, we were immediately captured by the beauty of the place. It was a relatively mild day, so the cold was not a problem, until the wind picked up. We quickly learned that Antarctica is beautiful but hostile at the same time. We settled into our tents at Union Glacier and would be making the last preparations over the next few days before we would fly off to 89 degrees, which is where our expedition would really begin. Again, this would be weather-depending, but we had plenty of things to do, so we didn't mind: prepare the pulka, practise our skiing, sort out our food, and go mountain biking if we had spare time as the weather was great.

The Ilyushin 76 is a Russian military
plane that can fit a truck inside.

We were very excited to learn that a number of famous explorers would come to Union Glacier that day. Firstly Robert Swan OBE, the first person in the world to walk to both the North Pole and the

South Pole. I had seen him once before at one of his lectures, so instantly I recognized him as his eyes are very distinctive. They're permanently altered from dark blue to light grey, having been exposed to the hole in the ozone layer on the South Pole in 1986. I was inspired when he told me he was training for his next expedition: skiing to the South Pole, together with his son Barney, without using any fossil fuels, only renewables. This had never been done before.

The big tents made our stay at Union Glacier relatively comfortable.

We were very honoured to also meet Sir Ranulph Fiennes, the world's greatest living explorer. He had just climbed Mount Vinson, the highest mountain on Antarctica. He explained that he was climbing the Seven Summits as that would make him the first person in the world to have climbed the highest mountain on each continent and to have crossed both polar caps. He had called this project the Global Reach Challenge. Sir Ranulph explained a number of Norwegians

Many Worlds to Conquer

Rob Swan and I promoting responsible investing,
which we are both passionate about.

were trying to do the same thing, but none of them had crossed the Arctic yet, something which is increasingly difficult given the open leads one encounters these days. He joked that climate change might not be such a bad thing after all!

At Union Glacier I also saw David Hamilton, who had guided Sir Ranulph to the summit of Mount Vinson, and Dave Hahn, who had also just guided on the mountain. They are two of the best mountain guides in the world. Dave Hahn even holds the world record for summiting Everest fifteen times. I had climbed with Dave Hahn on Denali in 2011, so it was great to see him again after five years.

Unfortunately we were told we would not be flying out the next day to our starting spot at 89 degrees, but the good news was that we were now able to attend an impromptu presentation by Sir Ranulph,

Ranulph Fiennes (left) telling us about his successful climb of Mount Vinson.

David, and Dave on their expeditions that Friday evening. Rarely will one see these kinds of legends together in one room, in this case the mess tent. It was a real privilege to see all these world-famous explorers together, especially as they decided to do a joint presentation on a topic that captures every explorer's imagination, Mount Everest. I was not going to miss that!

They managed to show some photos on a screen, and what a treat it was. Sir Ranulph told us he climbed the mountain in 2009, which was on his third attempt, and hence became the oldest Briton to stand on top of the highest mountain in the world at the age of 65. With typical self-deprecating humour, he said it was amazing what one can do with a bus pass these days!

David Hamilton had climbed the mountain a number of times and explained what it's like to go through the Khumbu Icefall at 5,500 metres, just above base camp. In short, it's absolutely nerve-racking: crevasses with a depth of almost 50 metres, which could swallow a building, overhanging seracs of 10 metres, which would kill you instantly if they were to fall on you, and of course avalanches that would sweep you away before you even knew it. And if you survive, you will reach the death zone at 8,000 metres. In the death zone your body starts to break down as it is so deprived of oxygen. You only have a third of the oxygen you have at sea level, which is something that a human being cannot survive. Even with an oxygen mask you won't last more than twenty-four hours.

But nothing beat the stories of Dave Hahn, in particular his discovery of the body of George Mallory. You might remember Mallory was the climber who was asked in 1924 why he wanted to climb Mount Everest and quipped, "Because it's there." There is one thing to mention about what happened after that comment. George Mallory went off to climb Everest with Sandy Irvine later that year and made good progress. They were spotted close to the summit by a fellow expedition member through his binoculars but then disappeared from view, never to be seen again. Until seventy-five years later.

Dave Hahn was leading the Mallory and Irvine Search Expedition in 1999. The team were looking for their bodies and had a good idea where Irvine had disappeared on the mountain based on his ice axe, which was found in 1933. What's known for sure about the Mallory and Irvine summit attempt is that they got close before they disappeared from view and perished afterwards. But no one knows whether they made it to the summit and met their deaths on the way down or whether they perished on the way up and didn't make it. This is the biggest unsolved mystery in the history of mountaineering.

The only thing that would give conclusive evidence of whether they were the first people to stand on top of Mount Everest would be the Kodak camera that Mallory and Irvine were carrying with them.

The Kodak company had confirmed they would be able to develop the photos, in the event the camera was found, as the cold would have preserved the film. It was reasonable to assume that Irvine had been carrying the camera, as Mallory was a celebrity who would have liked to have his pictures taken by Irvine, rather than the other way around. So finding the body of Irvine would be crucial to solving the mystery. And Dave and team knew where to look.

You can imagine the excitement of Dave when he discovered a body at 8,300 metres in the proximity of where Irvine's ice axe had been found. When he and his crew approached the snow-swept body that was exposed because of the windy and dry weather, he couldn't wait to find out whether Irvine had the camera on him or not. What surprised them was to find out that Irvine was wearing Mallory's clothes, as the shirt had a label which read *G. Mallory*. Only then did the team's oxygen-deprived brains realize this wasn't Irvine; this was the body of George Mallory, the most famous climber in the history of mountaineering!

Mallory's right leg was broken, which indicated that he must have fallen. The snapped rope around his waist also confirmed that. Furthermore his goggles were found in his pocket, which means that he was still alive when the sun had gone down. Just by taking a look at the remains of Mallory's body, you could see that he had been an incredibly strong human being. He definitely would have been able to make it, especially as he and Irvine had oxygen with them, even though it was in a primitive form. The weather had also been good. As Mallory had written in his diary that morning, "Perfect weather for the job."

But the most important question remains unanswered as the search team didn't find the body of Irvine with the camera he was carrying. Did they or did they not make it to the summit of Mount Everest?

All these stories were truly fascinating. Our expedition was already a special experience, although our own adventure had yet to begin! After this presentation we were more than ready for our own expedition.

We tried out our tents just outside of Union Glacier.

On Saturday we finally got the go-ahead to fly out to 89 degrees. It was a three-hour flight in a DC-3 (a Twin Otter), and when we landed on the ice it was about –30°C, which was cold but bearable. My left foot immediately started getting cold when I got out of the plane as my Baffin boots were a bit too small. The issue with my left foot is that it is slightly bigger than my right foot. And with double socks in boots that are too small, you are restricting your blood circulation. We didn't go skiing; we just set up camp, melted water, and prepared our first dinner. The dried spaghetti Bolognese tasted a lot better than I had expected. Although we would have liked to go out and start skiing that day, we were happy that the expedition had now truly started. Because it was twenty-four-hour daylight, we struggled to get some sleep, but that didn't bother us too much. We were looking forward to getting going on Sunday morning.

The sun didn't set during our expedition.

I wanted to try different systems to keep my foot warm. Obviously double socks didn't work, so I tried a single thin sock, which would probably be cold but comfortable. I added a chemical foot warmer that I attached to the sole of my foot. I'm not particularly keen on foot or hand warmers as it is often hit-and-miss with them. Sometimes they work, and sometimes they don't. But in this instance it worked, and I was very relieved. I was also hoping that my face mask wouldn't cause my goggles to fog up, which it didn't because I had cut out a big hole where my mouth was. Obviously I did risk freezing my lips, but it was a risk I had to take as I wouldn't be able to ski if I couldn't see anything.

It became apparent early on that Willem had a problem with his back. He started to take muscle relaxer pills. We only skied 12 kilometres on Sunday, and the overall distance was 120 kilometres, but our guide Christian Styve told us not to worry about that too much at this stage. The mood was good. But when we were about to set up camp, we heard a big rumble in the distance.

This was very odd as we couldn't see anything coming towards us on the ice or in the sky. But the sound was clearly coming our way. And it was getting louder and louder. "What is that?" I asked Christian. He didn't respond. At first I thought it was a very low plane flying over, but we had clear blue skies and there was no plane to be seen anywhere.

Then Serge asked, "Did you feel that?" I thought, *Feel what?* But then I did too. The ice that we were standing on had started to shake. By now we were all standing still, having nothing to hold on to, as the noise got louder and the trembling of the ice got worse. We all looked at each other in fear and were just waiting for what was going to happen.

The noise was coming our way and got louder and louder. By now the ice was shaking so violently that I was about to fall over. I had never been forced to contend with an earthquake before, but my mother grew up in San Francisco and had experienced her fair share of earth tremors. Her advice was always to stand in a door opening when the ground starts to shake, as it's normally the safest place. Obviously that was going to be difficult here on Antarctica. I felt helpless. There was nothing I could do if the ice were to open up and swallow us.

But then, like lightning underground, the rumble shot by and the noise disappeared. The ice got steadier, and we slowly got our breath back. "Christian, what the hell was that?" I asked.

He responded in a very matter-of-fact way, "Oh, that was just the ice cracking deep down, which happens when ice moves at great depths. Nothing to be worried about."

Then Willem said, "Well, don't you think this was an earthquake? Don't you think the ice could have cracked open and swallowed us?" Christian didn't seem fazed about it and said the ice never opens up. I wanted to ask him how he thought those crevasses had come about that we saw only an hour earlier, but then I thought I should maybe just shut up and be happy we were still alive.

Over dinner we talked about our plans for the rest of the expedition and came to the conclusion that we needed to do many more kilometres a day if we wanted to complete the expedition in a week, in time for our scheduled pickup.

We had a good day of skiing on the Monday and did 17 kilometres (from 9 a.m. until 4 p.m.). Willem's back really started becoming a problem, and we were not sure he would be able to make it. Furthermore, we heard on our daily call with Union Glacier (over satellite phone) that the next day we would have heavy winds and the temperature would drop to -50°C. We were strongly advised to stay in the tent for the whole day. Obviously that was the only sensible option, but we were running out of time to complete our Last Degree before our scheduled pickup. We still had more than 90 kilometres to go, and it was not clear what the plan was from here. This was not very uplifting, as most of the team members didn't have the flexibility to stretch the expedition. They needed to be home with their families for Christmas. Hopefully we would be able to make good progress on Tuesday, Wednesday, Thursday, and Friday, and with a big push the last day we would stand on the South Pole on Saturday. But everything had to come together perfectly, and everyone would need to push a lot harder, including Willem. I was getting worried. We were all coughing because of the cold, dry air, and Willem was suffering badly.

The absence of pressure ridges on the South Pole makes skiing a lot easier than on the North Pole.

Through the night the temperature started dropping, and when we woke up on Tuesday morning it was indeed bloody cold. We realized that staying in the tent would mean we would have to significantly increase our daily mileage for the rest of our expedition. But it was the sensible thing to do. After some discussion, the decision was taken to pack up camp anyway and start skiing. Although I didn't feel comfortable with this decision, I went along and immediately knew that we were in for a very tough day. The visibility was very low, the temperature was -50°C, and the winds were incredibly strong. Any piece of skin that was uncovered would immediately freeze, so we all double-checked each other's face masks to make sure that no skin was exposed. Willem had it tough from the moment we set off. He was slow, his back was hurting real bad, and he couldn't breathe properly. We all hoped things would improve along the way, but it all deteriorated very quickly. After half an hour we had to stop. He couldn't move anymore, and his breathing had become a real problem. We immediately set up camp again to get Willem in a sleeping bag and give him something warm to drink. Christian didn't know either what was the matter with Willem, but it was obvious he couldn't go on and needed an emergency evacuation. We were in the middle of nowhere. How long would it take for a plane to get here? And would it be able to land in these horribly windy conditions? The situation looked very dire, and our expedition would surely be over because of this delay that no one had foreseen.

When Christian called to Union Glacier to ask for an emergency evacuation, he was told that we were lucky because ALE had a DC-3 at the South Pole at that very moment. And he also heard that another team that was only a couple of kilometres away from us had also asked for an emergency rescue for two of their people. They too had ignored the advice from Union Glacier and had gone out to make some mileage. But would that plane be able to take off, and would it be able to land close to us?

After two hours our tent was opened by the emergency doctor, which took us totally by surprise. Where had she come from all of a sudden? The wind was blowing so hard that we hadn't even heard the plane when it landed next to our tents! Observing Willem's coughing and breathing, the doctor's diagnosis was HAPE, high-altitude pulmonary edema, which is life-threatening as fluid starts to accumulate in the lungs. It's a major cause of death among high-altitude mountaineers if not treated immediately. Although we were certainly not at Himalayan altitude levels, the impact of the elevation level of the South Pole is more severe because of the lower air density. To be more specific, the altitude of 2,835 metres here is the equivalent of 4,500 metres at the equator. Willem needed to be evacuated immediately. How lucky we were that there was a plane already on the South Pole and that it had been able to land in these conditions. This could have gone horribly wrong.

We were very sad to have lost Willem on our adventure, but his unfortunate departure gave us a minor consolation as we now could increase our daily distance. And something to cheer us up was the fact that it was Serge's birthday. He turned 50, which we celebrated with cheese and salami and a bottle of cognac that I had taken with me for this special occasion. Our spirits were now rising, but we all realized that going out again in this weather would be foolish as it was deteriorating even further, so we decided that we would stay in our tents for the rest of the day. It gave me the opportunity to attend to a big blister that was developing on my left foot and to check how many foot and hand warmers I still had with me. I would surely need them all in these brutal conditions. Over dinner we discussed the events of the day, and we concluded that had we stopped an hour later, the plane would not have been able to land, which would have put Willem's life in serious danger. We went to bed early as we had a long day ahead of us on Wednesday.

In the morning the weather was still poor, but it was good enough to go out again. Luckily it steadily improved throughout the day, and in the afternoon the temperature had risen to a balmy -15°C. We did 20 kilometres, despite the sastrugi we encountered, which are snow ridges formed by wind erosion. This distance was a record. We were now only 70 kilometres away from the pole. Christian also heard that our pickup from the South Pole was now on Sunday rather than Saturday which gave us a contingency day! The flight from Union Glacier back to Punta Arenas was scheduled for Monday or Tuesday, so we would still be back in time for Christmas!

Remko and I with Serge in the middle,
celebrating his 50th birthday in style!

In the evening we heard from the doctor, who told us that Willem had not suffered from HAPE but a double pneumonia. This was good news, but Willem was advised to leave Antarctica to recover and the flight out would take him another five days, he'd been told. I can only imagine how he must have felt.

On Thursday we had a similar day with very cold temperatures in the morning, but we got some sunshine in the afternoon. This allowed us to set a new record of 21 kilometres, and our spirits were high. By now we had less than 50 kilometres to go with Friday, Saturday, and possibly Sunday morning ahead of us with the pickup scheduled on

Sunday afternoon. But we wanted to get there as soon as possible as the weather is very unpredictable. If we pushed hard, we would be standing on the South Pole by Saturday evening. So was the hope. Christian had suffered frostnip on his face as he hadn't worn a face mask that day. He got worried and found out the hard way that putting Vaseline on your face doesn't give you enough protection. It might have worked in Norway but not on the South Pole, not even for someone who had been in the Norwegian special forces. It fortunately didn't develop into a frostbite, but it was a good reminder of the hostile conditions and that we certainly shouldn't let our guard down, although we were getting closer and closer.

The next day we all felt strong but didn't manage to break the distance record from the day before. Christian had started to develop blisters, and over dinner he admitted that he had underestimated the intensity of this expedition. I had worn my face mask all day, but I felt a windburn just below my left eye, which I somehow got even though my face had been fully covered. I decided to put Vaseline on that spot the next day, underneath my face mask. We only had 30 kilometres to go, and with focus and determination we would be able to make it in one day.

We set off early on Saturday morning, and it was another cold day indeed. I wore all the protection I had and constantly had my hood up, which caused my goggles to steam up. This was terrible, as every twenty minutes I had to stop to clear my goggles. Serge told me to leave the hood down like he did. I didn't want my windburn to develop into frostbite and played it safe, but the team wasn't too happy with me as I held the group up.

After skiing more than ten hours, we finally saw the Amundsen–Scott Research Station in the distance, and there and then we knew we were going to achieve our goal.

I was wearing a double face mask to prevent frostbite.

When it started to sink in that soon we would stand on the South Pole, I felt elated, as not many people get to experience that. What an honour this was. I realized what the intrepid explorers before us must have gone through to get to this point. At 8.30 p.m. we were all humbled to finally stand on the South Pole. What a feeling. We dedicated our success to Marc Cornelissen as he had been the inspiration for our expedition in the first place. We quickly took pictures and then had a big dinner in the mess tent that ALE had set up. Only about twenty-five people a year ski the Last Degree South Pole, so we were all thankful for this truly remarkable experience.

On Sunday we were lucky enough to be invited to the Amundsen–Scott Research Station. It is a data collection centre set up in 1956 by the Americans to study the geophysics of the South Pole and hosts around 150 people at its peak. Important data is collected here regarding climate change, the ozone layer, and glacial movements.

Very honoured to stand in the footsteps of
Scott and Amundsen on the South Pole.

We were impressed by the people who were willing to work there, as far removed as possible from civilization just for the sake of science.

In the afternoon the Twin Otter was able to pick us up. At 11.45 p.m. we landed back at Union Glacier. The next day at breakfast we saw that Serge had developed a black spot on his right cheek and two black spots on his left cheek. This did not look good. The base camp doctor came up to our table and asked Serge to visit her after breakfast, as she told him this clearly was frostbite and needed to be looked after. She gave him aloe vera and told him the skin would simply rejuvenate but would remain sensitive to the cold for the rest of his life. That evening we had a big celebratory dinner, and we all immensely enjoyed this last night we would have at Antarctica. The Il-76 would fly us back to Punta the next day, just in time for our flight on Wednesday to Europe to be with our families at Christmas. What a way to finish the year!

This One's for Marc!

Chapter 9

ESCAPE FROM ALCATRAZ

Alcatraz is situated on a small rocky island in the Bay of San Francisco. It used to be a disciplinary barrack and was acquired by the US Department of Justice in 1933. It opened its doors in 1934 as a federal prison. And not just any federal prison. This one was designed to put away those criminals who caused too much trouble in normal prisons. Its guests were the most notorious murderers, bank robbers, and gangsters in the country. Their hosts, the prison guards, were highly trained and the toughest around, also known as "iron men". During the twenty-nine years that Alcatraz served as a federal penitentiary, no one is said to have escaped. In total, thirty-six inmates did attempt to, but they were all caught alive, were shot dead, or presumably drowned. Alcatraz closed its doors as a prison in 1963 and is now open to the public.

Alcatraz was designed to make escaping impossible. The cell doors of this maximum security prison are made of hardened steel. Even if a prisoner were to get out of his cell, he would still have to penetrate three additional security gates before he'd get to the yard. And the yard was enclosed and watched by the armed prison guards, who were all trigger-happy. After that he would still have to climb the outer wall that was covered with barbed wire and surveyed by six watchtowers. From there on he would need to get down a rock wall of about 15 metres to get to the water. Once he got to the water, he would be met by the guards who were patrolling the surrounding waters in their powerboats. And if a prisoner managed to pass them, he would still have to deal with ice-cold water, the strong currents, and the sharks in the San Francisco Bay to get to the nearest land. In other words, escaping from Alcatraz was impossible.

But somehow, in December 1962, an inmate by the name of John Paul Scott did manage to escape the unescapable prison, and when he got to the water, he swam towards the closest land point in the San Francisco Bay but was swept out by the current all the way to Fort Point, near the Golden Gate Bridge. When he got ashore, he was completely hypothermic and couldn't move anymore. A young boy spotted him and called the police, after which Scott was escorted back to the prison he had just escaped from.

And then there is the famous case of bank robbers Frank Morris, Clarence Anglin, and John Anglin. They also managed to break out of the prison in 1962, and their story is one of the most ingenious escapes ever. At night, over a period of six months, they had gradually scraped a hole in the wall underneath the sinks in their cells. For that they used stolen spoons and discarded blades. During the day they managed to conceal those holes with cardboard and plaster, but at night they would climb through them to get to a utility corridor which led onto the roof of their cell block. There they would work quietly on makeshift floating devices fashioned from stolen raincoats.

The guards never noticed their absence at night as the three prisoners used handmade dummy heads with hair that they had taken from the floor of the barbershop to fool the prison guards and make them believe they were sleeping quietly in their cells. This trick worked well all the way to the end.

On the morning of 12 June 1962, the guards noticed that these three men hadn't left their cells and were still sleeping. They went in to wake them up only to find these dummy heads. It was then they realized that these men had escaped during the night. The guards searched their cells and found the holes that had been dug in the walls underneath the sinks. The three inmates had climbed through the openings to get to the utility tunnel, then the ventilation shaft, after which they climbed two 4-metre barbed wire fences and then managed to get to the water to set out for Angel Island, 3 kilometres north of Alcatraz, never to be seen again.

During the massive search that followed, some remnants of raincoat materials were found on Angel Island beach, but the men were nowhere to be found. The FBI concluded that the escapees most likely drowned because of the cold water and the strong currents. But their bodies were never recovered, and it remains a mystery up until today whether they actually made it to Angel Island or not.

The story of this escape led to the famous Clint Eastwood movie *Escape from Alcatraz*, which I watched over and over again when I was a kid. My brother and I loved to see how Clint Eastwood in the role of Frank Morris misled the nasty prison guards. The final scene of the movie is shot on Angel Island beach, where the FBI director is told by his subordinate, "Tides were mild and the fog light last night. They left with lights out and had a nine-and-a-half-hour head start. I wonder if they made it." The FBI director's resolute answer is, "They drowned," to which the subordinate replies, "Yes, sir."

I went to San Francisco for the first time in 2003. My visit would not have been complete without a visit to Alcatraz, and when I did set foot on "the Rock", I immediately got a creepy feeling. There is still a lot of barbed wire, and the towers where the guards had been watching the prisoners looked intimidating. When I walked into the prison, I could almost sense the dread of these notorious criminals. Alcatraz was not a friendly place, and probably it was never meant to be. I went inside one of the cells, which were very small, and could only imagine what it must have felt like to be locked up there. I would have done anything to get out. Just like Frank Morris and the Anglin brothers.

In 2012 I had started a new job, which is when I visited San Francisco again. Our offices were in a fifty-two-storey skyscraper, and from high up I had a great view of the Bay Area. I saw Alcatraz right in the middle, and I could see Angel Island in the distance. Somehow I just thought, *I can swim that.*

Alcatraz with Angel Island in the distance.
I should be able to swim that!

When I started researching this, I found that some coaches do indeed help swimmers to "escape from Alcatraz", but only in a southerly direction to San Francisco, which is a 2-kilometre swim, and not north to Angel Island, which is a 3-kilometre swim. And the latter is what I wanted because that was the swim that the three prisoners presumably had undertaken. I also learned that you have to be a very competent open water swimmer, but I didn't even know how to swim with a crawl stroke. Obviously I shouldn't even consider escaping from Alcatraz, but I wasn't ready to give up and managed to get hold of Pedro Ordenes, an Alcatraz crossing legend.

Pedro explained that swimming Alcatraz brings a number of challenges that you have to be very aware of. "TJ," he said, "the water temperature ranges from 10 to 16 degrees Celsius, which is very cold. Water from the Sacramento and San Joaquin rivers, but more importantly the Sierra Nevada mountain range, flows through the Golden Gate. To a large extent this is melted glacier water, which explains the cold temperature. Have you ever been in water that cold?" Melted glacier water seemed cold indeed.

"No, I don't think I have. What could I compare it to?"

Pedro had to think and then said, "You've probably been to a sauna before, after which you jump into a cold water tub to improve your circulation. That's exactly how cold the water in the Bay Area is." *Yikes,* I thought.

He then continued: "You should also know that these waters connect in the San Francisco Bay Area with the Pacific Ocean, which creates strong currents around Alcatraz. Every second, five million gallons of water run underneath the one-mile-wide Golden Gate Bridge, which is pretty much next to Alcatraz. Furthermore, you have to be aware of the marine life. Seals might look cute, but sea lions definitely don't and aren't, and they are constantly there. And they are pretty sizeable."

I actually knew that as I had gone to Pier 39 on Fisherman's Wharf when I was in San Francisco for the first time. The sea lions congregate there in large numbers, and I remembered they looked very big indeed. In fact, they can weigh up to 500 kilos!

"Lastly there are five species of sharks in the Bay Area, but the good news is that great whites generally don't come into the bay because of the brackish water. They mainly reside in the ocean, near Farallon Islands, about thirty-five kilometres away from San Francisco." Somehow I didn't find this reassuring at all.

Pedro urged me to take swimming lessons, get experience with open water swimming, and call him again in a year's time. And so I did. I went to Emile Bitar, the swim coach at the Royal Automobile Club in London, and he patiently taught me about bilateral breathing, streamlining, hand angle, arm entry, body rotation, kicking, body positioning, and sighting in open water. During one clinic, once my technique started to develop, he asked me to put on a wetsuit, as I was going to wear one for my Alcatraz swim, and do some laps again. This was like magic. The neoprene in the wetsuit improves your buoyancy dramatically, which made everything so much easier! He also told me that swimming in a pool is much tougher than swimming in the sea, as salt water also gives you more buoyancy. Obviously a wetsuit would also help against the cold, as the water that gets into your suit will warm up to your body temperature and keep you comfortable. This all significantly boosted my confidence. I couldn't wait to get some real open water experience.

In London the Serpentine Lido in Hyde Park offered me the perfect opportunity to swim 100-metre laps outdoors, which I started in May. It was a good way to get used to swimming in a wetsuit in fifteen-degree water, which would probably be the same temperature as the San Francisco water in August/ September when it would be at its warmest. Swimming in the Serpentine Lido was a very different way to enjoy Hyde Park. I absolutely loved it. The one thing that was a bit of an issue was the fact that I couldn't see where I was going. In a swimming pool you can simply follow the line on the bottom, which you can see clearly. Here I couldn't see anything, as the water is so dark, and to swim in a straight line from one side to the other is almost impossible. Sighting while swimming is absolutely necessary in open water. Obviously I still had to try swimming in choppy water with currents, but now I felt ready to try that too.

During my 2013 summer holidays I trained in the open sea on many occasions and found out that this also adds a number of challenges: waves, currents, and winds. But at the end of July I felt ready for Alcatraz!

In early August I talked to Pedro again and told him I had trained for a year. He invited me to join him for a one-day Alcatraz swim clinic, which I could fortunately fit into my schedule as I had to be in San Francisco for my work again. I immediately accepted his offer and was excited but also a bit apprehensive to finally try out the water there. I had only talked over the phone with Pedro, but when I met him in person, he immediately put me at ease. He's a very gentle man who has a tremendous amount of open water swimming experience and patiently helped me prepare my first open water experience in the Bay Area. I normally don't take a wetsuit when I go on a business trip, and this time it was no different. Pedro was kind enough to help me rent one in San Francisco.

He then took me and three others on his boat to Alcatraz. It was scary to jump off into the murky water. What would be lurking underneath? I didn't even want to think about it. Once in the water, your body gets an initial shock reaction because of the cold, but after a while you get over it and you get your breathing back. Pedro told us to swim towards the Golden Gate Bridge, against the currents.

I swam as hard as I could to counter the current, and with my wetsuit on I didn't get hypothermic. The biggest challenge was mental. Knowing that there are plenty of big animals in the water beneath me, which I couldn't see, was very scary. I had done a bit of homework on this and knew that the sediment in the water reduced visibility, which would make it more difficult for sharks to hunt. And I had read that the high level of fresh water on the surface, as it is melted glacier water, would make it difficult for sharks to breathe. So even if they were swimming in the deep water below me, they wouldn't be able to get to me. But somehow all this textbook theory wasn't comforting at all. I was bloody frightened.

The currents were incredibly strong, which became clear when Pedro told us to stop and come back onto the boat. Although we had been swimming in the direction of the Golden Gate Bridge for ten minutes, we had ended up hundreds of metres in the opposite direction and were being swept back to Alcatraz! Clearly the currents are a big challenge too. It's very easy to misjudge them as you can't see them.

We swam at different places in the bay, and every time I got onto the boat I was relieved, only to jump into the water a couple of minutes later. But this whole day had given me even more confidence in my open water swimming abilities. The waters had been rough, but I had managed to hold my own and felt strong when swimming.

I was very pleased when Pedro told me at the end of that day that he would be willing to help me escape from Alcatraz! We would coordinate further details when I got back to London.

He knew that I wanted to swim north to Angel Island, rather than south to San Francisco, and that is not a swim he normally organizes. But this year he would make an exception, not in the least because the prestigious America's Cup was held that year in San Francisco and the Bay Area was closed off for these elite sailors. That was great for me to hear. I'd gotten lucky! We agreed that Saturday, 31 August, would be the day, weather permitting. I booked my flight to arrive two days before.

When I got to San Francisco that Thursday, I participated in a group swim in Aquatic Park in the early evening, but my performance was lousy. To start off with, I lost my goggles in the murky water and had to go back to shore to get my reserve pair. I also got cramps in my feet, probably because of the dehydration from the twelve-hour flight from London. Furthermore I was pretty tired because of the jet leg and hence slow. Most of the others were excellent swimmers, so I clearly stood out like a sore thumb. When I had had my swim clinic a couple of weeks earlier, my fellow swimmers were a number of 60-year old ladies, which had been much better for building my confidence!

Watching the America's Cup was a good distraction.

On Friday my wife and I took it easy and went to see the boats racing for the America's Cup along the coastline. It was incredible to see these catamarans lifted out of the water to create as little drag as possible. They achieve speeds of up to 75 kilometres per hour. I had never seen any boat sail so fast. It was spectacular to watch. This cup was particularly interesting as this was sailing history in the making:

Oracle Team USA was trailing Team New Zealand 1–8, but in one of sport's greatest comebacks, they managed to win 9–8. I felt very fortunate to have seen part of that spectacular race.

That Friday evening I went to bed on time as I had to be ready early on Saturday morning. My wife joined me for a big breakfast at 5.30 a.m. You need to eat a lot, which will help against potential seasickness. Obviously I needed a lot of carbohydrates to give me energy, so I had plenty of toast, with eggs and bacon for protein. I also made sure to have two bananas as they are rich in potassium, which helps prevent cramps, as did my electrolyte-fuelled drinks. I drank two litres as cramping would be a serious issue because you can't relax and stretch your feet or legs in rough open waters. And I wouldn't be allowed to go back onto the boat; that's against the rules. Finally I double-checked I had everything with me: earplugs to avoid seasickness, my swim goggles, my wetsuit, and my swim cap. In fact, I had two caps, as you lose about 25 percent of your body heat through your head, so I wanted to cover it as well as I could.

I met Pedro at the harbour straight after breakfast. He wanted to leave for Alcatraz as soon as possible. The winds are the weakest and currents the least problematic early in the morning. He had mentioned to a couple of people that I was going to swim the "true" escape from Alcatraz, which had enticed some other swimmers to come along. I felt honoured because you could already see that before they jumped into the water, these were elite swimmers, unlike me. They had V-shaped bodies with very wide shoulders, and some of them didn't even bother to wear a wetsuit!

I took a good look at Alcatraz before I jumped into the water. I also tried to spot the mountain on Angel Island which Pedro had told me to swim towards as soon as I hit the water. Halfway I would need to change direction and swim towards the windmill on the right of the mountain, and then the current would push me

I'm about to jump in. Anxious and excited at the same time.

towards Angel Island beach, that Frank Morris and the Anglin brothers had possibly reached in 1962. It all sounded straightforward, but the visibility was very low because of heavy fog. This could also turn into a safety issue because it becomes difficult to spot troubled swimmers in the water. Fortunately my cap was bright green so I would be easy to find in the water in case of emergency.

After I jumped in, it took me a couple of minutes to get into a good rhythm. My body needed to get accustomed to the cold as I had expected, but the bigger issue was that I couldn't see the bloody mountain anymore from the water. Sighting in choppy waters is very difficult indeed, especially on a foggy day. Pedro had told me I would be able to enjoy the scenic view of the Golden Gate Bridge and the city of Sausalito, but nope, I didn't see any of that either once I was in the water.

As I'd been first in I couldn't follow anyone, so I started to swim towards what I thought would be the middle of Angel Island, in the direction of the mountain. After a couple of minutes my nerves had subsided and I was swimming steadily, only to find out that a number of the other swimmers were not swimming in my direction, nor towards the windmill. They were going east, straight for the beach. This worried me as these were the elite swimmers who had come to join me on this swim. They couldn't have gotten it wrong; *surely my direction would be off*, I thought. Fortunately I chose to ignore them, as ten minutes later they were being swept out by the current, which they were now fighting ferociously. The support boat had seen this happening and was able to pick them up after it became clear, thirty minutes later, that they had no chance of fighting the current. They were then repositioned so they could continue their swim. This served as an indicator of how treacherous these waters can be, even for elite swimmers.

Halfway through my swim I started to focus on the windmill that was now clearly in view, and I could also see the beach in the distance. It was tempting to aim for it straight, but I was very happy to stay the course, as I noticed I was now being slowly pushed in an easterly direction by the current, towards the beach. At about 500 metres from the shore I could see that the sand strip was quite wide, and I knew that it was looking good now. There was no way I was going to miss this beach and be pushed out past Angel Island, so I started to relax and take it all in. I felt incredibly proud a quarter of an hour later to finally set foot on the beach that I had seen many times over in the Clint Eastwood movie.

This had been a tough swim in the cold water, with strong currents and undoubtedly lots of sea life swimming underneath me. But I had made it, and it showed that an escape from Alcatraz can be done under the right conditions, which was also the case on the night of 11 June 1962, when Frank Morris and the Anglin brothers escaped. I am absolutely convinced that the three inmates reached Angel Island too. If I can do it, they can do it. And let's be honest, they had a much better incentive than I had!

On the boat back after my successful escape
from Alcatraz. My wife Kym and her sister
Lesli had come along to support me.

Another reason why I was happy to have successfully escaped from Alcatraz was that it allowed me to raise funds for my nephew Felix. The prior year, on 3 October 2012 to be exact, I received a call from my twin brother Boudewijn, who was living in New York at the time. I had just gone for a swim at the Royal Automobile Club and was walking along the Mall on my way home. I was approaching Buckingham Palace, which looked beautiful at sunset, which is when my brother rang. I was looking forward to a nice chat, especially as he had just celebrated his eighth wedding anniversary the prior day. I was keen to hear how his boys, 2-year-old Felix and 6-month-old Julius, were doing. My wife and I don't have children, so I spend as much time with my nephews as I can. They also see something in me that reminds them of their father, maybe because we're twins. I'm incredibly proud to be their godfather, and we have great fun together.

My brother sounded very distressed. I immediately stopped walking. It was early afternoon in New York, but I knew he wasn't in the office as I heard a lot of traffic in the background. "Where are you?" I asked.

"I'm in the taxi on my way back from Mount Sinai Hospital, and we just received very bad news. Felix has been diagnosed with type 1 diabetes." I was somewhat familiar with the condition and what it entailed following my adventures with Adrian during the New York City Marathon and the Mont Blanc. I asked whether that meant he needed injections. "Yes he does. We've just been shown in hospital how to inject him. I've never felt as bad as I did this morning when I injected him for the first time, as he couldn't stop crying." Mind you, Felix was only 2 years old and had no understanding of why his father had to put a needle in him. "The worst part is that he needs six injections per day. For the rest of his life. When we come home we have to do the second."

I knew that Felix had not been well for a couple of weeks. I had seen him in August when we celebrated my mother's eightieth birthday in the Netherlands. Felix had not been his usual self. He was grumpy, and I assumed he was tired because of the jet lag he had incurred from the transatlantic flight. It also could have been that he wasn't receiving enough attention because his baby brother Julius had been born a couple of months earlier. But what do I know about children? I chose to ignore it.

When my brother and Anne, his wife, returned to New York, there was not much improvement. Felix had started to look paler and paler, had developed more bags under his eyes, and was starting to lose weight.

Anne didn't trust it and went to the doctor with Felix on 2 October, the date of her and my brother's wedding anniversary. The doctor got suspicious when he heard that Felix had to pee more often than usual, and he performed a number of tests. He would know the outcome in the evening and said that Anne and my brother had to come back to the hospital first thing the next morning to hear the results. This obviously worried Anne. The moment she got home, she immediately started her research on the internet to be well prepared, just in case. Their worst nightmare came true when they managed to get hold of the endocrinologist from Mount Sinai hospital that evening. This was not the wedding anniversary gift they had hoped for.

It is still unclear what the cause is of type 1 diabetes, whether it is genetic or not, and what triggers it. Some people are diagnosed at a very early age, like Felix at the age of 2, but some people only become diabetic when they are fully adult. Whatever the cause or the trigger, Felix will be living with diabetes for the rest of his life. One can only imagine the stress this brings to my brother and his wife. I wanted to do anything to help and decided to raise money for the Juvenile Diabetes Research Foundation (JDRF), which focuses exclusively on type 1 diabetes. It seemed the obvious thing to do. When I decided to "escape from Alcatraz" in aid of the JDRF, I hadn't actually heard much about the charity. I had just found them on the internet, but my brother knew it was a highly regarded global charity. I was very impressed when the organization contacted me during my fundraising campaign and wanted to know if there was anything they could do to help me. This was a very professional charity indeed, which gave me even more conviction to support them. The charity angle made this adventure a worthy cause and was a great way to get to know the organization and its research initiatives a bit better. After they received the funds I'd raised, they awarded me a Gold Medal Fundraising Award and asked me to sit on their advisory board, which I was very honoured to do. It was the start of a long-term partnership with the JDRF.

In New York with my nephew Felix and
Rufus the Bear, the JDRF mascot.

Chapter 10

CROSSING THE ENGLISH CHANNEL

This chapter is an adaptation of my article that was published in SUPboarder Magazine *on 11 March 2016*

The first person to cross the English Channel on a stand-up paddleboard was Laird Hamilton in 1990. And the crossing has been completed several times since, either solo or as a team, and usually by paddlers with a lot of experience in open water. TJ Halbertsma decided to take on the busiest shipping lane in the world after paddling for less than one year and with little open water experience. Don't worry, he had some company! Here's TJ's account of his training in Hawaii, with wind, waves, currents, and heavily armed officials!

"Passeports, s'il vous plaît," hollered the machine-gun-toting French coast guard from the deck of his boat, flanked by nine other stern armed officials. I knew I had to bring my passport, but having my paperwork checked as I bobbed in the middle of the English Channel? That came as a surprise.

We had just entered French waters, and the coast guard was on high alert. Little did we know that Calais had just been inundated with refugees and no one was to enter or leave France without the right paperwork. Although my passport was perfectly valid, the coast guard didn't seem convinced they should let me continue my journey. Had I paddled this far for nothing?!

Fortunately I was given the chance to explain what I was doing before they sent me back. "This is a paddleboard. I'm trying to cross the English Channel ... and I'm doing it for fun," I politely informed them. They seemed a bit confused. I wasn't sure whether it was my story or my French. After discussing the situation, they must have concluded I was not a migrant, and they let me continue my Channel crossing. Oh, the relief—I had trained so hard to get to this point, and my goal was now within reach.

Cross-channel training can be fun.

It all started a year earlier, in 2014. I was in Hawaii visiting my parents-in-law, and we were talking about paddleboarding, which I hadn't really picked up yet. Over dinner I heard about Laird Hamilton, the best big wave surfer in the world and all-round waterman, who was the first person to ever cross the English Channel on a paddleboard. And only a handful of people have ever done it.

I was captivated by the idea and knew it warranted further exploration. I couldn't find much information on it and realized that very few people had followed in Laird's wake, which made it even more exciting. This was something that had to be done.

I had my first training session in January 2015 when I was in Laguna, California. I booked an instructor, and once out at sea we focused on the basics—the stance, the balance, and the strokes. After an hour I started to get the hang of it. The rhythm of the strokes, the freedom of the water, it really made me feel alive. With dolphins leaping to my left and whales breaching 200 yards to my right, I was so overwhelmed that I immediately lost my balance and fell in. But boy, the whole afternoon had been such a great experience. I was hooked.

Back in London, I reached out to Paul from Active 360, who organizes stand-up paddleboarding (SUP) on the Thames. I signed up for their sessions a couple of times a week, did a course in San Pietro, Sardinia, over Easter, and then wanted to really test my mettle by taking part in the 40tude SUP Marathon that June.

The 42 kilometres from Shepperton to Putney would be about the same distance as the Channel, so if I could nail that, I would probably be able to do the crossing too. OK, so there wouldn't be cargo ships to negotiate, strong currents to navigate, and buffeting crosswinds to contend with, but surely this would be some indicator of my competence?

I did well in the event and felt strong throughout, which really boosted my confidence. By now my technique was pretty good. I had done plenty of core training in the gym and had been able to avoid injuries.

But there was one thing missing from the marathon that would make all the difference on the English Channel: waves. Thankfully, a month before my planned crossing I was back in Hawaii, where I had plenty of swell to contend with on my SUP, which was great for getting my sea legs.

The last couple of training sessions I did in Hastings to get used to the water temperature. Hastings is not Hawaii, and the water was pretty cold with strong winds of up to 20 knots. But after a while I noticed that falling in was actually a pretty efficient way to cool down. I also learned that taking a break on your board to eat and drink is not a good idea given the strong currents. You have to eat and drink while you paddle, and that takes some practice. Lastly Hastings provided me with the opportunity to try out a couple of different boards. A longer board will create more speed, but it shouldn't be too long, as that would make it difficult to navigate the waves. The trade-off for the width of the board is similar. If the board is very narrow, you will go faster, but that will go at the expense of stability, which you need much more at sea than on the Thames. I ultimately chose a Starboard Freeride XL 12'2" × 32' and felt ready for the crossing. For safety reasons I had organized a support boat, captained by a guy named Will (from Full Throttle Boat Charters) whom I had never met. He told me to be on standby the whole week of 3–7 August. On the Tuesday, I got the call to be in Rye the following morning at six o'clock. We were on!

When we met, we didn't have time for a long introduction. It was now or never. Will explained to me, "The currents are now in our favour but will turn against us in a couple of hours. You will not beat them, so if we haven't reached the shipping lane by midmorning, you won't be able to make it to France. We'd better get going!" I knew

I would have to paddle fast and work hard to get to the shipping lane, 15 kilometres away, before the currents would start working against me. And this shipping lane, with six hundred tankers and two hundred ferries passing every day, was also something that I didn't look forward to. But that was not something to think about now. So, off I went.

Once on my board, I found it pretty daunting to paddle out on to the sea with nothing in sight but open ocean. I was also told not to look back because the white cliffs of Dover will stay in sight forever and you'll get the feeling that you're not progressing. Of course, I did look back, and my God, it took hours before they finally disappeared.

After four hours of very hard work, I finally got to the shipping lane and was utterly spent. The closer you get to the world's busiest maritime motorway, the bigger the oil tankers become and the higher the waves they create. At one point I couldn't even see these huge vessels anymore as the waves they were creating were actually higher than me on my board! This was the last thing I needed, as I was now feeling the strain of hours of hard paddling. But this is where my training in Hawaii came in very handy. I was able to handle the waves without falling off or having to kneel on my board. Currents had indeed picked up, and the last hour had been absolutely gruelling.

Once I entered the shipping lane, I had to get my board out of the water and get onto the support boat, as you're not actually allowed to cross the shipping lane on a paddleboard. To compensate for the 5 kilometres in the boat, you start in Rye rather than Dover so you still paddle the same distance. The moment the support boat got to the other side of the shipping lane, I could start paddling again. This is where the French waters start and the grilling from the coast guard began.

Giving way to supertankers and bulk carriers is part of the challenge.

After they had decided to allow me to paddle on, they wished me well with a very friendly "bonne continuation". Unfortunately, Will was told to stay behind to have his boat searched. Obviously they didn't find any refugees on board, but they didn't seem intent on letting him continue. I was now totally alone in the middle of the sea. I crossed my fingers and paddled on.

After an hour I started to see sailboats and realized I was getting close. When the shore was finally coming into sight, Will had caught up with me, and his cheers of encouragement came just at the right time. I tried to comprehend that I actually might make it and was able to start enjoying my crossing during the last stretch. And there it was, the coast of Normandy. It was within my reach.

When I finally got to the beach, I wasn't allowed to go ashore, and we had to go straight back to England. But it didn't bother me. I had just crossed the Channel in five hours and forty-five minutes and could hardly believe it! I had counted on seven or eight hours, so this was well beyond expectations. This wasn't bad at all if you consider the world record is five hours and nine minutes.

TJ happy to have finished his challenge.

When I started to relax in the support boat on the way back, I was aching all over. Paddleboarding is a full body workout after all. I also found out that Will and the reserve skipper, Henk, had placed a bet as I saw some cash exchanging hands. I was curious and asked them what the bet was about. Will looked happy and said, "Before we set off, we bet on the number of times you would fall off your board. Henk said more than fifty and I said fewer. You fell off only three times. Well done, and thank you!"

Back in Rye, I was surprised to see a large number of people waiting for me in the harbour. When I was offloading my board, one of the men asked me how far I had gone out. I explained that I had crossed the Channel. Again he wanted to know how far I had gotten. "Well," I said, "all the way to France."

"I'm a paddleboard instructor. That can't be done," he chirped.

With a smile on my face, I replied, "Young man, I just did!" Will had to chuckle.

Back in London that evening, I took a hot shower and was glad that everything had gone according to plan. It had been a long day. To celebrate I opened a bottle of wine—French, obviously—and I raised a glass to the coast guards who had been kind enough to let my dream become a reality. And what a journey it had been!

Epilogue

Three years later, in the summer of 2018, my wife and I were lucky to have the opportunity to spend a couple of days with the legendary Laird Hamilton and his wife Gabby Reece at their house in Malibu. During those days it became quite clear that he is an extraordinary human being, very driven in everything he does. I saw that he trains harder than anyone I've ever seen, and he's a big believer in getting out of your comfort zone to achieve your goals.

That became particularly apparent when we were doing a workout with weights in his garden. He said that anyone can do that on land, but doing it under water adds a different dimension. I thought he was kidding, but he obviously wasn't when he told me to jump into his pool with those weights! Next thing I knew, I was lifting weights 4 metres under water, but once I filled up with CO_2, I felt the urge to breathe and came to the surface. He then explained that you have enough oxygen in your blood to stay under water for a couple of minutes. But it is your mind that has this built-in survival mechanism that says you need to get to the surface and breathe. "Try to control your mind," he said. "Stay calm."

He was teaching me about the mental toughness that is required when you're getting out of your comfort zone. After a number of breathing exercises that were aimed at calming the mind, he told me to try again, and I was indeed getting better and better, staying under water for longer and longer.

Under water pool training really got me out of my comfort zone.

He told me he went through this process himself when he was growing up in Hawaii, carrying big stones under water as he wanted to become a big wave surfer. He needed to be able to hold his breath for a long time in the event he fell off his board riding a big wave and got dragged under water. He's now able to stay under water for more than five minutes! Laird is known as the best big wave surfer of all time, and it started to become apparent what level of commitment that had required.

Obviously I also told him about my English Channel crossing. The first question he asked was, "Did you use a paddle?" That seemed a bit of a strange question as paddleboarding is normally done with a paddle! I told him I very much did, after which he told me he hadn't on that legendary first crossing.

"So what did you use?" I asked.

"Nothing, just my hands!" Clearly, there's only one Laird Hamilton.

Surf's up! With Laird Hamilton in Malibu.

Ain't No Mountain High Enough

On a beautiful summer day in 2013, I was walking up to the Almageller Hut in Switzerland to climb the south-east ridge of the Weissmiess the following day. On our way to the hut we passed a group of 12-year-old children from Belgium. I was surprised to see them there, as mountaineering is not something that normally attracts young kids. Fortunately their guide explained they wouldn't climb the mountain; they would only do the trek to the hut, stay the night, and then go down the next day. I was equally surprised that these kids didn't seem to have any difficulties with the long trek up to the hut at 2,894 metres. In fact, they really enjoyed it and were in great spirits when they arrived at the hut. They all seemed proud of their achievement and had big smiles on their faces. As mountain huts are very simple with only basic facilities, I wouldn't describe a night in a mountain hut as special. But this time, with the voices of these happy kids, and the sight of fifty pairs of children's shoes next to all those big, heavy mountaineering boots, the stay was very cheery.

I had to think of Felix and wondered whether he would be able to do something similar when he grew up. I remembered my Mont Blanc climb with Adrian only too well and had learned from that adventure that one has to be extra careful in the mountains if one has diabetes. But then again, if you manage your diabetes well, there is no reason why you wouldn't be able to climb a mountain. Children with diabetes can sometimes hold back a bit, and their parents can sometimes be overprotective, so very few ever get the chance to climb a mountain unless someone else encourages them to do so. I then came up with the idea to climb a mountain with diabetic kids. Making it to a summit would be such a great confidence builder for them, but clearly we had to be careful. I had seen how happy these Belgian children were after their climb up to the hut, and nothing would give me greater joy than organizing a similar trip for kids with diabetes.

When I got back to London, I discussed the idea with the Juvenile Diabetes Research Foundation, and asked whether they would be able to help me organize the event with the aim to raise money for the charity again. They loved the idea and immediately offered to help me as much as required. They would screen their database to see who might be interested, and they also would organize a doctor to come along.

With the support from the JDRF, I knew I would be able to pull this off. I was excited! I still had to pick a mountain and soon decided it would be Ben Nevis in Scotland, the highest mountain in the United Kingdom. It's not a particularly difficult climb, and you can trek up to the 1,345-metre summit in a couple of hours. But it is still a strenuous undertaking, and the weather can always be a challenge in Scotland. Clearly this would get youngsters out of their comfort zone, especially given that they had diabetes. Summiting the highest mountain in the country would feel like a real achievement to them. And to me too, if I could help them do it. I decided to call this project Ain't No Mountain High Enough.

The date would be Sunday, 23 August 2015, just before the kids would go back to school. The charity had arranged to have GP Grant McCallum join us for medical supervision. After careful consideration, Grant had informed us that the minimum age should be 14 years old, which meant that my nephew Felix wouldn't be able to come along. This was a disappointment, but I wasn't going to argue; safety comes first.

We had eight "climbers" signed up, mostly from Scotland, and agreed to have dinner the night before so we could all get to know each other. In the end we had a pretty large group together as all the parents had come along too. Most of them had pretty thick Scottish accents. When they heard my Dutch accent, they politely asked whether I could understand what they were saying. "Aye, of course ah dae," I replied, which broke the ice that evening.

It struck me how little they all knew about diabetes. I was told that the dissemination of information on diabetes and insulin itself is just not very good in Scotland. Grant was able to explain where you can get insulin in case of emergency, and he also talked about the latest developments in the treatment of diabetes. Everyone was particularly interested to hear that one can now get an insulin pump as an alternative to injections. This small device will deliver insulin via a little tube by remote control, and you only need to replace the tube every couple of days. In other words, you no longer have to inject six times a day. I have seen this with Felix; insulin pumps are life-changing. Most of the group had never even heard of their existence.

Adele, one of the girls who would be climbing with us, explained she would go up Ben Nevis for her mother, who had just suffered a diabetes-related stroke which she probably could have prevented. She explained that it was difficult for her mother to get insulin in their hometown, so sometimes she just didn't inject. Her mother was there too and didn't say much. I got the feeling she felt embarrassed that she had suffered a stroke. She was probably not even fifty years old. It was an emotional moment. Adele said, "Mum, I'm going to climb Ben Nevis for you as I want to show you how important it is for me that you take good care of yourself."

Her mother was clearly moved and said, "But it wasn't until tonight that I knew what treatment options there are. I'm so sorry. I'm going to take care of myself, and I'll make sure it will never happen again." This evening already made Ain't No Mountain High Enough a very special project, and I felt very thankful that Grant had been at that dinner.

But I was most taken aback by Jono's story, who hadn't attended the dinner. I only met him the next day just before we set off. He didn't seem to be the most social person. He worked in construction, had a big tattoo on his neck, and had been diagnosed with diabetes at the age of 5. He said he had ignored it when he was younger, as he "wanted to be like the other lads". Unfortunately he had paid the price already as he had suffered a diabetes-related heart attack only six months earlier. In addition he smoked and had a big ulcer on his foot, which is not a good sign, as 80 per cent of foot amputations start with an ulcer. Jono was 29 years old, but only now he realized how serious a condition he had. I was impressed that he had joined our group, as he was suffering from the moment we set off. His ulcer was really bothering him. In fact, he was limping.

Because I knew that we would probably have some slower climbers in the group, I had arranged two guides. If a climber had to go down, then one guide would be able to facilitate that while the other guide would be able to continue up with the group.

Fortunately the weather had cleared when we started. After two hours into our trek, the sun began to come through. This was a real lucky break as the night before it had been storming and raining. It did raise everyone's spirits and certainly helped Jono continue. We were doing well, but Jono was struggling as we continued. Grant said he would be there for him throughout the climb, but boy, Jono didn't want to have any of it. "Leave me alone. I don't need your help!"

The higher we got, the more the winds became an issue. With gusts up to 120 kilometres per hour, we sometimes had to stand still. We just couldn't move forward. But after four hours we did make it to the summit of Ben Nevis, the highest mountain in the UK. And we all made it, including Jono. I went up to him to congratulate him, and I saw that he was crying. I was surprised but also humbled. "TJ, I've looked at this mountain every day of my life, and I never imagined I would be able to stand on the summit." I had seen how much he had suffered physically with the ulcer on his foot and having just recovered from his heart attack. I could hardly apprehend how difficult it must have been for him mentally, too, to commit himself to this challenge. I gave him a hug.

I told him, "I'm so proud of you that you did not let diabetes beat you. And please also know that our climb already raised £20,000 for JDRF!"

He looked baffled. "I didn't think anyone cared about me or my diabetes. Thank you for making this happen!" He was a totally different person from the Jono I had met only hours before. Standing on the summit was a dream come true for him. I was very honoured to share that moment with him.

For me, that moment really captured what Ain't No Mountain High Enough was about. Get out of your comfort zone and live your dream. This trek up Ben Nevis was also a special experience for me as I had had the opportunity to make a difference. I don't have children of my own, but I realized on that mountaintop with those youngsters that there are also other ways to give back and feel meaningful. Mountaineering can have a purpose after all.

Throughout this book I have talked about my exploits, and I hope you found them inspiring. Sometimes people invite me to speak about my adventures, and as you probably know by now, I try to tell everyone that you can do anything you want as long as you set your mind to it. And when I get the question of who inspires me, I explain that the real heroes, the ones who inspire me, are people like Jono. And Adrian with whom I had climbed the Mont Blanc. They truly show us all that there Ain't No Mountain High Enough!

All the proceeds of this book will be donated to the JDRF.

Lightning Source UK Ltd.
Milton Keynes UK
UKHW012346260220
359388UK00001B/8/J